W9-CUE-216

The Essential Guide to Telecommunications

Annabel Z. Dodd

To join a Prentice Hall PTR Internet mailing list,
point to: **http://www.prenhall.com/mail_lists/**

Prentice Hall PTR
Upper Saddle River, NJ 07458

ISBN 0-13-259011-5

90000

9 780132 590112

Library of Congress Cataloging-in-Publication Data

Dodd, Annabel.
    The essential guide to telecommunications / Annabel Dodd.
       p.   cm.
    Includes bibliographical references and index.
    ISBN 0-13-259011-5
    1. Telecommunications.   II. Title.
   TK5101.D54   1997   97-35218
   384--dc21   CIP

Editorial/production supervision: *Lisa Iarkowski*
Composition: *Lisa Iarkowski and Beth Sturla*
Cover design director: *Jerry Votta*
Cover design: *Bruce Kenselaar*
Manufacturing manager: *Alexis R. Heydt*
Acquisitions editor: *Michael Meehan*
Marketing manager: *Stephen Solomon*

Prentice Hall books are widely used by corporations and government agencies for training, marketing, and resale.
The publisher offers discounts on this book when ordered in bulk quantities. For more
information, contact:

        Corporate Sales Department
        Phone: 800-382-3419
        Fax: 201-236-7141
        E-mail: corpsales@prenhall.com

Or write: Prentice Hall PTR
             Corp. Sales Dept.
             One Lake Street
             Upper Saddle River, NJ 07458

Printed in the United States of America
10 9 8 7 6 5 4

ISBN 0-13-259011-5

Prentice-Hall International (UK) Limited, *London*
Prentice-Hall of Australia Pty. Limited, *Sydney*
Prentice-Hall Canada Inc., *Toronto*
Prentice-Hall Hispanoamericana, S.A., *Mexico*
Prentice-Hall of India Private Limited, *New Delhi*
Prentice-Hall of Japan, Inc., *Tokyo*
Simon & Schuster Asia Pte. Ltd., *Singapore*
Editora Prentice-Hall do Brasil, Ltda., *Rio de Janeiro*

*This book is dedicated to my parents Harry L. Zalc
and my late mother Sara Zalc who taught me the meaning
of hard work and perseverance. To my husband Bob,
who believed in my ability through every facet of my career;
and to my three daughters, Judith, Nancy and Laura
and my son-in-law Stephen Concannon.*

Contents

Acknowledgments

I would like to thank the many people who took the time to speak with me for this book. Numerous individuals at telecommunications companies talked with me about telecommunications technologies. Staff at General Instruments, Motorola Corporation, Cisco Systems, Inc., Community Newspaper Company, Chrysalis-ITS, SBC, NYNEX, UUNet Technologies, MCI, Ascend Communications, Netscape Communications, Nextel, GTE, MFS, Teleport and Qwest Communications are among the companies who provided me with information. Numerous students who attended the classes I teach at Northeastern University's State-of-the Art Program helped me more than they can know. Myra Collard at Cisco Systems, Nancy Passavant at GTE and Michael Doiron at CellularOne come to mind. They generously provided me with contacts of knowledgeable people in their organizations. All of my students brought up issues in class that form the basis of each chapter in my book.

Colin Crowell, Congressman Markey's aide for Telecommunications was an inestimable help on the background and main features of the Telecommunications Act of 1996. Dr. Ronald O. Brown, President of Brown Consulting in Melrose Massachusetts provided technical details on local area networks for Chapter 1. Walt Tetschner, President of Tern Systems in Acton, Massachusetts, provided invaluable background and statistics about the wireless industry. Scott A. Helmers, Director, The Harvard Computing Group, Inc., in Westford, Massachusetts, assisted with information about the Internet. He also pointed out important topics to cover.

Thanks also to Rhonda Htoo, MIS Manager, Goldhirsh Group, for her insight on local-area networking applications and to Peter Barnes for background information on Northern Telecom's pioneering use of digital technology in central offices. Peter also shared with me his own experiences using cable modems. Ruth Winett of Winett Associates in Framingham, Massachusetts, pointed out newspaper articles about privacy on the Internet and sources of statistics on cellular service. Joel Winett of BGS Systems spoke with me about email attachments. Barbara Noyce of Motorola ISG generously supplied contact names at Motorola and timely advice. Peter McGowan, of m-g marketing communications helped me tremendously with his comments about the book cover design and insights on commerce on the Internet. In addition, Nancy Stimac of Community Newspaper Company helped me by reading my initial chapters and encouraging me to finish the book. She also provided me interesting material on how her department is using the Internet to sell Classified Ads.

My Executive Editor at Prentice Hall, Michael E. Meehan, gave me much needed advice on the structure of each chapter and encouragement to complete the book. I couldn't have written this book without the support and help of all of these people.

Most of all, I would like to thank my husband. Bob provided a layman's critique of the book. He took the time to read each chapter multiple times and to offer insightful comments. No easy task. His help was invaluable and filled with common sense and intelligence.

Preface

Telecommunications changes do not occur in isolation. They evolve over time in conjunction with changes in business, technological advances, and changes in regulatory and economic conditions. Key components of this book are explanations of how telecommunications technologies evolved, how they are used and their impact on business and society. Descriptions are included on how organizations, telecommunications vendors and telephone companies use and market telecommunications services. Intertwined with high-level technical explanations are examples of the role played by telecommunications in our safety, security and national economy.

A "sorting out" of industry "players" and vendor types is presented to provide readers with an understanding of the impact of the Telecommunications Act of 1996. The Telecommunications Act of 1996, itself, is examined in light of its effect on consumers, commercial organizations and telephone carriers. Technologies important in competition for local calling, high-capacity communications and Internet access are explained.

This book is intended for non-technical people working in the field of telecommunications and for people responsible for the administration of telecommunications services for their organizations. They include regulatory staff, salespeople, marketing personnel, process designers, human resources people, project managers, telecommunications managers and high-level administrators.

The Essential Guide to Telecommunications explores the language and significance of important telecommunications technologies. It is not intended to be a deeply technical book. Rather, it is an overview of technologies and an explanation of the structure of the industry.

The Essential Guide to Telecommunications starts out with an explanation of fundamental concepts so that readers will have a basis for understanding more complex, new telecommunications services. It explores the structure of the industry, local competition, the Telecommunications Act of 1996, the Internet and wireless services.

Intertwined with explanations of technology are examples of applications and histori-cal highlights. How the industry evolved and how the technology changed are also explored. The stories and explanations that accompany the technical details are key to the book.

PART 1

Fundamentals

CHAPTER 1

Basic Concepts

T his chapter defines basic telecommunications terms. Understanding terminology is important for staff who work with technical employees and customers. According to my students at Northeastern University, learning the telecommunications "language" gives them the tools to know which questions to ask. For example, a student in the finance department of her organization no longer feels intimidated when engineers ask for huge amounts of money for technical projects. She now has the vocabulary and confidence to ask pointed questions about proposed projects. Another student works in public relations and needs to understand concepts explained by engineers. These and other students have improved their effectiveness with co-workers, engineers and customers by understanding and interpreting technical vocabulary.

Terms such as analog, digital and bandwidth are used in the context of services that touch the everyday work experiences of professionals. In particular, in their everyday lives, people access the Internet and work from home. They are finding more and more that older, analog services are not adequate for these functions. They don't have the speed, meaning bandwidth, to download Internet pages or corporate files fast enough. Organizations where part of the workforce telecommutes want to provide remote workers fast enough telephone lines to access remote files so that they have the tools to work effectively. To get the speed they need, they often deploy digital telephone lines, such as ISDN, at both workers' homes and corporate locations. The quality of telecommunications services has a direct bearing on worker productivity.

Another important concept is protocols. Protocols are similar to etiquette between like computers. Just as etiquette spells out who shakes hands first, how people greet each other and rules for how guests should leave parties, protocols spell out the order in which computers take turns transmitting and how long computers should wait before they terminate a transmission. Protocols handle functions such as error correction, error detection and file transmissions in a common manner so that computers can "talk" to each other. A personal computer sends data to another personal computer using a protocol such as Kermit, a protocol designed for personal computers.

Computers, printers and devices from different vendors also need to be able to send information such as electronic mail and attachments across networks. This is the role of architectures. Architectures tie computers and peripherals together into a coherent whole. Layers within architectures have protocols that define functions such as routing, error checking and addressing. The architecture is the umbrella under which the protocols and devices communicate with each other.

Computers located in firms' offices are physically connected together over local area networks, LANs. LANs are located within a building or in a campus environment. LANs are made up of computers, printers, scanners and shared devices such as modems, video conferencing units and facsimile units. LANs are connected to other LANs over MANs, metropolitan area networks, and WANs, wide area networks. The growing number of devices and peripherals on LANs is adding congestion to data networks. Workers encounter network congestion when there are delays in transmission and receipt of, for example, email and database look-ups. This chapter reviews why there is congestion on local area networks and ways companies can eliminate this congestion.

One solution to traffic jams on wide area networks is the use of multiplexing. Multiplexing enables multiple devices to share one telephone line. For example, T-1 provides 24 paths on one telephone line. Newer multiplexing schemes add even more capacity. T-3 provides 672 channels on one telephone line. These multiplexing schemes provide private and non-profit organizations with ways to carry increasing amounts of data, video and imaging traffic between sites. T-3 is an important way for large call centers, such as airlines, to handle large volumes of incoming calls.

Another way to add capacity for applications such as graphics, x-ray images and video is the use of compression. Compression squeezes large amounts of data into smaller sizes. It is analogous to putting data into a corset. As a matter of fact, the availability of affordable video conferencing systems is made possible by advances in compression. Compression makes the video images "fit" onto slower speed telephone lines than those required before advances in compression. Before advances in compression were made, the high-speed telephone lines needed for video conferencing were prohibitively expensive.

Finally, grasping fundamental terminology creates a basis for learning about advanced telecommunications services. Understanding fundamental concepts, such as digital, analog, bandwidth, compression, protocols, codes and bits, provides a basis for comprehending new technologies such as ATM, SONET, ADSL and wireless services. These new technologies are changing the way Americans do business, spawning new telecommunications services and creating a smaller, linked, worldwide community.

Analog and Digital

The public telephone network was originally designed for voice telephone calls. When the telephone was invented in 1876, it was used to transmit speech. The telegraph, invented in 1840, was used for short text messages. Spoken words are transmitted as analog sound waves. People speak in an analog format, waves. Telephone calls were transmitted in an analog form until the late 1960s. While much of the public telephone network is now digital, there are still many analog services in use and portions of the telephone network are analog. The "plain old telephones," POTs, that fit into standard telephone jacks are analog instruments. Current TV signals, most telephone lines from homes to the nearest telephone company equipment and cable TV drops, the cabling portions from subscribers to their nearest telephone pole, are analog.

As more people use their computers to communicate, and as calling volumes increase, the analog format, designed for lower volumes of voice traffic, is proving inefficient. Digital signals are faster, have more capacity and contain fewer errors than analog waves.

High-speed telecommunications signals sent on ISDN service, within computers, via fiber optic lines and between most telephone company offices, are digital. With the exception of current TV and portions of cable TV wiring, analog services are used for slow-speed transmissions. Analog services are mainly POTs lines used by residential and small business customers.

Analog Signals

Speed on Analog Services

Analog signals move down telephone lines as electromagnetic waves. Their speed is expressed in frequency. Frequency refers to the number of times per second that a wave oscillates or swings back and forth in a complete cycle from its starting point to its end point. A complete cycle, as illustrated in Figure 1.1, occurs when a wave starts at a zero point of voltage, goes to the highest positive part of the wave, down to the negative voltage portion and then back to zero. The higher the speed or frequency, the more complete cycles of a wave are completed in a period of time. This speed or frequency is stated in hertz (Hz). For example, a wave that oscillates or swings back and forth ten times per second has a speed of ten hertz or cycles per second.

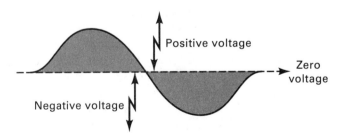

One cycle looks like a "resting" letter S

Figure 1.1   One cycle of an analog wave, one hertz.

Analog services, such as voice, radio and TV signals, travel within a specified range of frequencies. For example, voice travels in the 300 to 3300 Hz range. The speed that a service travels at is determined by subtracting the lower range from the higher range. Thus, the speed that voice travels at within the public network is 3000 hertz (3300 minus 300) also expressed as Hz or cycles per second.

Speeds for analog services are expressed in abbreviated forms. For example, thousands of cycles per second is expressed as kilohertz (KHz)and millions of cycles per second is expressed as megahertz (MHz). Analog transmissions take place in enclosed media such as coaxial cable, cable TV and on copper wires used for home telephone services. They are also transmitted via "open" media such as microwave, home wireless telephones and cellular car phones. Particular services are carried at predefined frequencies. Examples of analog frequencies are:

- Kilohertz or KHz = thousands of cycles per second
  Voice is carried in the frequency range of .3 KHz to 3.3 KHz, or 3000 KHz.
- Megahertz or MHz = millions of cycles per second
  Analog cable TV signals are carried in the frequency range of 54 MHz to 750 MHz.
- Gigahertz or GHz = billions of cycles per second
  Most analog microwave towers operate at between 2 and 12 GHz.

The 3000 cycles allocated to each conversation in the public network is slow for digital computers when they communicate on analog lines via modems. Modems, which enable digital computers and facsimile machines to communicate over analog telephone lines, have methods of overcoming some of the speed limitations in the public, analog portion of the network. (See Chapter 3 for information about modems.)

Impairments on Analog Services

Sending an analog telephone signal is analogous to sending water through a pipe. Rushing water loses force as it travels through a pipe. The further it travels in the pipe, the more force it loses and the weaker it becomes. Similarly, an analog signal weakens as it travels over distances whether it is sent over copper, coaxial cable or through the air as a radio or microwave signal. The signal meets resistance in the media (copper, coaxial cable, air) over which it is sent, which causes the signal to fade or weaken. In voice conversation, a voice may sound softer. In addition to becoming weaker, the analog signal picks up electrical interference, or "noise" on the line. Power lines, lights and electric machinery all cause noise in the form of electrical energy to be present along with the analog signal. In voice conversations, noise on analog lines is heard as static.

To overcome resistance and boost the signal, an analog wave is periodically strengthened with a device called an amplifier. Amplifying a weakened analog signal is not without problems. In analog services, the amplifier that strengthens the signal cannot tell the difference between the electrical energy present in the form of noise and the actual transmitted voice or data. Thus, the noise as well as the signal is amplified. In a voice telephone call, people hear static in the background when this happens. However, they can generally still understand what is being said. When noise on data transmissions is amplified, the noise may cause errors in the transmission. For example, on transmitted financial data, the received sales figures might be $300,000 and the sent information $3 million.

Digital Signals

Digital signals have the following advantages over analog:

- Higher speeds.
- Clearer voice quality.
- Fewer errors.
- Less complex peripheral equipment required.

Clearer Voice, Fewer Errors

Instead of waves, digital signals are transmitted in the form of binary bits. The Second College Edition of *The American Heritage Dictionary* defines binary as being composed of two parts. In telecommunications, the term binary refers to the fact that there are only two values for transmitted voice and data bits, on and off. On bits are depicted as ones, positive voltages, and off bits are depicted as zeroes, negative voltages. The fact that digital transmissions are only on or off is one reason why digital services are more accurate and clearer for voice. Digital signals can be recreated more reliably. It is more complex to recreate a wave that can have multiple forms than a bit that is either an on or off pulse.

Both analog and digital signals are subject to impairments. They both decrease in volume over distance, fade and are susceptible to interference, such as static. However, digital signals can be "repaired" better than analog signals. Figure 1.2 illustrates that when a digital signal loses strength and fades over distance, equipment on the line to regenerate the signal knows that each bit is either a one or zero and recreates it. Noise, or static, is discarded. The noise is not, as in an analog signal in Figure 1.2, regenerated. To illustrate the clarity of digital services, people in Maryland who first used digital wireless telephones rather than analog cellular service commented on the improvement in voice clarity over analog cellular service.

In addition to clarity, digital signals have fewer errors. In analog transmission, where noise is amplified, receiving equipment may interpret the amplified signal as an information bit. People using modems to transmit data often receive garbled data. In digital transmissions, where noise is discarded, garbling occurs less frequently, thus there are fewer errors in the transmission.

Digital Television—An Example of Digital Transmission to Enhance Clarity

Analog television standards were approved by the FCC in 1941 for black-and-white television. (Widespread television introduction was delayed by World War II.) Color TV standards set by the NTSC, National Television Standards Committee, were

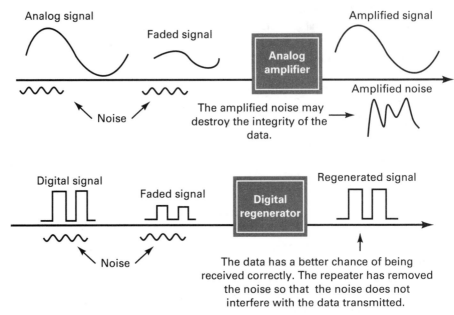

Figure 1.2   Noise amplified on analog lines; eliminated on digital service.

approved in 1954. As people with analog broadcast television know, "snow" and "ghosts" are frequently present along with the television images. TVs located far from broadcast antennas have the most problems with clarity. This is a function of analog signals fading or weakening. The "snow" present on TV screens is interference from noise on the television channel. In this case, the noise or interference has become stronger than the signal. The further from the broadcast antenna, the greater the amount of noise relative to the picture being transmitted.

A factor in improved picture quality with digital television is the elimination of noise. With digital television, error correction code is sent along with the TV signal. This additional 10% of error correction code provides digital TV with the same clarity 50 miles from an antenna as 5 miles from an antenna. The error correction code checks the signal and eliminates errors. The error correction code "corrects" the signal from within the TV receiver. Thus, the clarity of the digital signal is uniform throughout the range of the antenna.

Moreover, digital signals degrade or weaken less over distance than analog signals. A digital signal must travel further before it starts to weaken or fade. The transition in terms of quality from analog to digital television is analogous to the change in quality from analog audio tapes to digital compact discs, CDs. Digital TV will provide studio-quality audio and image on home screens. Networks will broadcast analog as well as digital transmissions until 2006, unless the FCC extends this deadline past 2006. At the end of the simulcasting term, analog frequency channels will be turned over to people who purchased the frequencies sold by the FCC at public auctions.

Digital Television—TVs Act Like PCs

High-definition digital television will allow broadcasters to transmit secondary, non-programming information, as well as television signals. A 20 megabit per second data channel has been set aside to bring information services such as weather forecasts, home automation, audio for audio's sake and stock quotes into homes. This ancillary channel will be used in conjunction with interactive, remote control devices. For instance, a user will be given the choice of downloading technical specifications, pricing and warranty notices in conjunction with a car commercial.

Just as personal computers manipulate bits in the form of word processing, spreadsheet and financial programs, digital televisions will be receiving and manipulating a stream of bits. *In essence, whether used by cable television or commercial broadcast television, digital television will send digital bits into peoples' homes. The bits will be audio, video or text images.* The TV receiver, or in the case of cable TV, a set top device, will act as a computer and manipulate the signals to be viewed on the home screen. In telecommunications, a bit is a bit whether the source is the Internet, corporations or entertainment services.

Higher Speeds and Reliability

In addition to clarity, digital transmissions are faster than analog transmissions. This is because digital signals are less complex to transmit. They are either on or off bits, whereas analog signals take the form of complex waves. Whereas the highest speed projected for analog modems is 56,000 bits per second when receiving data and 33,600 bits when sending data, ATM and SONET are digital services which run at Gigabit per second speeds.

Finally, digital service is more reliable than analog. Less equipment is required to boost the signal. Analog signals weaken and fade at shorter distances than digital signals. At every point that a signal fades, amplifiers or regenerators are required. Each amplifier is a place for a possible failure. For example, water can leak into a telephone company's manhole or the amplifier itself might fail. Organizations that use digital lines such as T-1 often experience only one or two brief failures in an entire year. High reliability results in lower maintenance costs for the telephone companies that support digital circuits.

Digital Telephone Company Equipment—Saving Money on Maintenance and Space

Prior to the 1960s, both the transmission of calls and equipment to route calls were analog. Beginning in the 1960s, calls were first carried in digital format on cabling between central offices with analog switches. It was cumbersome to connect digital call traffic to analog for processing by analog central office switches. Devices called channel banks were needed to convert digital signals to analog to be handled within the analog central offices and to convert analog central office signals to digital to be carried on digital coaxial cable running between central office toll switches. Converting to digital central

Digital Services in the Bell System

Digital technology was first implemented in the public network in 1962. It was implemented, not in routing calls (central office switches), but rather in the transmission of calls within the long distance portion of the AT&T network. Coaxial cable between the central offices first carried digital calls. Because the digital technology was faster and was capable of carrying higher volumes of calls than analog technology, digital service was implemented as a way to save money by decreasing the amount of cabling required to carry high volumes of traffic. Fewer copper or coaxial lines were needed to carry equal volumes of digital rather than analog traffic.

Northern Telecom introduced the first digital telephone system switch for routing calls in 1975. However, to cut its financial risk, it first introduced the switch as a customer premise switch rather than a central office switch. At that time, telephone systems installed on customer premises were highly profitable and it was felt that there was less financial risk in introducing a smaller digital telephone system for end-users, rather than a larger, more expensive telephone company central office switch.

Significant dates for digital services are:

1962: T-1 on two pairs of telephone cable carried 24 voice or data calls in digital format.

1975: The first digital telephone system (PBX), the Northern Telecom SL-1.

1976: AT&T's #4 ESS toll office switched calls between central offices.

1977: Northern Telecom's central office switch, DMS 10, was installed in Canada. It was not installed in the U.S. until 1981.

1982: AT&T's #5 ESS central office switched calls from central offices to local homes and businesses.

offices eliminated the requirement for this analog-to-digital and digital-to-analog conversion equipment. This saved telephone companies money on:

- *Maintenance* on channel banks for the analog-to-digital conversion, and vice versa.
- *Space* required in the central offices for channel banks.

Bandwidth—Measuring Capacity

In telecommunications, bandwidth refers to capacity. Bandwidth is expressed differently in analog and digital transmissions. Analog bandwidth is referred to in terms of hertz. The speed of a particular media, such as coaxial cable, is referred to in hertz. Hertz is a way of measuring the capacity or frequency of analog services. For example, someone might say coaxial cable has a bandwidth of 400 MHz; 400 MHz means four hundred million cycles per second. The speed of the cable can be stated as a frequency of 400 MHZ. The bandwidth of an analog service is the difference between the highest and lowest frequency within which the media carries traffic. Cabling that carries data between 200 MHz and 300 MHz has a bandwidth, or frequency, of 100 MHz. The higher the hertz, the greater the capacity or bandwidth.

On digital services such as ISDN, T-1, T-3, SONET and ATM, speed is stated in bits per second. Simply put, it is the number of bits that can be transmitted in one second. T-1 has a bandwidth of 1.54 million bits per second.

Bandwidth in terms of bits per second or hertz can be stated in many ways. Some of these include:

- Individual ISDN channels have a bandwidth of 64 thousand bits per second, 64 kilobits or 64 Kbps.
- T-1 circuits have a bandwidth of 1.54 million bits per second, 1.54 megabits or 1.54 Mbps.
- One version of ATM has the capacity for 622 million bits per second, 622 megabits, or 622 Mbps.
- Another version of ATM has the capacity for 13.22 billion bits per second, 13.22 Gigabits or 13.22 Gbps.
- Any speed above Gigabit is called terabit, which is ten to the twelfth; 10 terabits = 10,000,000,000,000.

Narrowband vs. Wideband—Slow and Fast

In addition to bits per second and hertz, speed is sometimes referred to as narrowband and wideband. Just as more water fits into a wide pipe, wideband lines carry more information than narrowband lines. The term narrowband refers to slower speed services and wideband refers to higher speed services. Again, digital speeds are expressed in bits per second and analog speeds are expressed in frequencies.

The definition of which technologies are wideband and which are narrowband differs within the industry, as can be seen in Table 1.1.

Table 1.1   Wideband and Narrowband Telecommunication Services

Narrowband	Wideband
T-1 at 1.54 Mbps 24 voice or data conversations on fiber optics, infrared, microwave or two pairs of wire.	**Broadcast TV services— uses 6 MHz per channel** Newer digital high-definition TV (HDTV) offers enhanced clarity over analog TV.
Analog telephone lines at 3000 Hz Plain old telephone service (POTs). Modems enable analog lines to carry data from digital computers.	**Cable TV (CATV) and Community antenna television at 700 MHz** Broadcasts local and satellite TV. Also available for data communications and access to the Internet.

Table 1.1   Wideband and Narrowband Telecommunication Services *(Continued)*

Narrowband	Wideband
BRI ISDN at 144 Kbps Two paths for voice or data, each at 64 Kbps. One path for signals at 16 Kbps.	***ATM- up to 622 Mbps, megabits*** A very high-speed service capable of sending voice, video and data.
	SONET—Up to 2.5 Gbps, Gigabits An optical interface for high-speed transmission. Used mainly in carrier and telco networks.
	T-3 at 44.7 Mbps, megabits (equivalent to 28 T-1 circuits) A way of transmitting 672 conversations over fiber optics or digital microwave.

Television and cable are carried at wideband speeds. Lines connecting telephone offices together use wideband services. Voice calls, video and data transported within carriers' networks are generally carried at wideband speeds. However, most traffic from central offices to individual homes and businesses are carried at the slower, narrowband speeds.

Protocols and Architectures

Protocols—A Common Language

Protocols allow like devices to communicate with each other. They provide a common language and set of rules. Devices communicate over the Internet using a suite of protocols called TCP/IP. For example, the IP, or Internet protocol portion of TCP/IP, allows portions of messages called datagrams to take different routes through the Internet. The datagrams are assembled into one message at the receiving end of the route. Other protocols enable communications among personal computers within an organization's building. The most common of these protocols is Ethernet. Personal computers use protocols such as Kermit, XModem and YModem. Apple's Mac computers can be connected to each other over the Apple Talk protocol.

Examples of protocol functions are:

- Who transmits first?
- In a network with many devices, how is it decided whose turn it is to send data?
- What is the structure of the addresses of devices such as computers?
- How is it determined if an error has occurred?
- How are errors fixed?

- If no one transmits, how long is the wait before disconnecting?
- If there is an error, does the entire transmission have to be resent or just the portion with the error?
- How is data packaged to be sent, e.g., one bit at a time or one block of bits at a time? How many bits are in each block? Should data be put into envelopes called packets?

Protocol structures have implications on speed and efficiency. The following protocols for personal computers illustrate this point:

- *Kermit:* Enables PCs to communicate with DEC minicomputers and IBM mainframe computers. A slow protocol.
- *Xmodem:* Enables a PC to send entire files, such as a word processing or spreadsheet document, to another PC or to the Internet. This protocol allows a PC to send one block or group of bits at a time and then waits until the recipient acknowledges that the data has been received correctly before it sends the next block of data.
- *Zmodem:* A newer, faster protocol. The sending PC does not have to wait for the receiving end to acknowledge that each group of data sent was received accurately. It keeps sending data until the receiving end sends a message indicating errors in the transmission. It then retransmits data starting with the group of bits that had errors.

Architectures—Communications Framework for Multiple Networks

Architectures tie dissimilar protocols together. Architectures are developed by standards bodies and dominant companies, like IBM. By the mid-1970s, IBM had sold its customers a variety of printers, terminals and mini- and mainframe computers. These devices communicated with each other by a variety of incompatible protocols. An architecture was developed by IBM to enable its devices to talk together. This architecture is called SNA, and it is *specific to IBM*.

During the same timeframe, an architecture was developed by a standards body, the International Standards Organization, ISO. This architecture is OSI, Open System Interconnection. OSI was developed to allow devices from *multiple vendors* to communicate with each other. It is an open architecture.

While OSI has not been widely implemented, it has had a deep influence on telecommunications. It laid the foundation for the concept of open communications among multiple manufacturers' devices. The basic concept of OSI is that of layering. Groups of functions are broken up into seven layers. These layers can be changed and developed without having to change any other layer. Both LANs and the Internet are based on concepts developed by the OSI for a layered architecture.

Layer 1 is the most basic layer, the physical layer. It defines the electrical interface (plugs) and type of media, for example, copper, wireless and fiber optics.

Layer 2 is the data link layer. LANs, networks within corporations, correspond to Layer 2 of the OSI model. They provide rules for error control and gaining access to the network.

Layer 3 is called the network layer. It has more complex rules for addresses and more error control than Layer 2. Communications *between* networks generally adhere to protocols corresponding to Layer 3 of the OSI. Layer 3 protocols are responsible for *routing* traffic between networks or sites.

The Internet suite of protocols, TCP/IP, corresponds to the functions in Layers 3 and 4 of the OSI model. These functions are addressing, error control and access to the network. The TCP/IP suite of protocols has provided a uniform way for diverse devices to speak to each other from all over the world. It was developed in the 1970s by the U.S. Department of Defense and was provided at no charge to end-users in its basic format. Having a readily available, standard protocol is a key ingredient in the spread of the Internet.

Compression and Multiplexing

Compression—Manipulating Data for More Capacity

Compression is comparable to a trash compactor. Just as a trash compactor makes trash smaller so that more can be stuffed into a garbage barrel, compression makes data smaller so that more information can be stuffed over telephone lines. It is a technique to get more capacity from telephone lines.

Modems—Using Compression to Get Higher Throughput

With compression, data to be transmitted is made smaller by removing white spaces and redundant images, and by abbreviating the most frequently appearing letters. For example, with facsimile, compression removes white spaces from pictures and only transmits the images. Modems use compression to achieve higher rates of transmitted information, or throughput. When modems equipped with compression transmit text, repeated words are abbreviated into smaller codes. For example, the letters E, T, O and I appear frequently in text. Compression will send shortened versions of these letters with 3 bits rather than the entire seven bits for the letters E, T, O and I. Thus, a page of text might be sent using 1600 bits rather than 2200 bits.

Modems use compression to send greater amounts of computer data in less time over analog lines. This increases throughput. For example, if a word processing file is ten pages long, compression that eliminates white spaces, redundant characters and abbreviates characters might compress the document to seven pages. Seven pages of data takes less time to transmit than ten pages. This is an example of compression increasing throughput, the amount of information sent through a line in a given amount of time. Telecommuters who access and send data to corporate locations often use modems equipped with compression to transmit files more quickly. Matching compression is needed at both the telecommuter's home and the corporate site (see Figure 1.3).

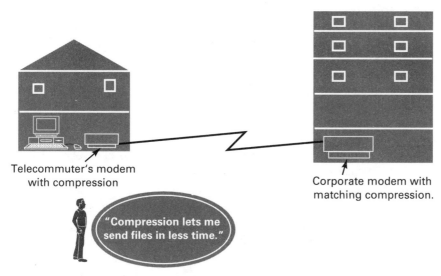

Figure 1.3 Compression in modems.

Video—Compression Made Video Conferencing Commercially Viable

In video, compression works by transmitting only the changed image, not the same image over and over. For example, in a videoconference meeting with a person who listens, nothing is transmitted after the initial image of the person listening until that person moves or speaks. Similarly, fixed objects such as walls, desks and background are not repeatedly transmitted. Another way video compression works is by not transmitting an entire image. For example, the device doing the compression, the coder, knows that discarding minor changes in the image won't distort the viewed image noticeably.

Improvements in the mid-1980s in video compression spawned the commercial availability of room-type video conference systems. It made it economical to use video by requiring less bandwidth, which translates into cheaper telephone lines. The older compression systems required a full T-1 for video. This inhibited the sales of room-type video systems until the late 1980s. New compression techniques in the 1980s from companies such as PictureTel required only 56 Kbps to 128 Kbps for acceptable picture quality.

Thus, video conferencing became affordable to a wide range of organizations. For example, instead of using a T-1 at hundreds of dollars per hour, organizations could use a service from someone such as MCI for as low as $14 per hour and still have acceptable video capability. New compression algorithms meant that slower speed digital lines were an acceptable choice for video meetings. A new industry boomed.

Compression and Standards—So Everyone Can Talk

There are many types of compression methods. Companies such as AT&T, Motorola, PictureTel and Compression Labs have all designed unique compression schemes using mathematical algorithms. A device called a codec, short for coder-decoder, encodes text, audio, video or image using a compression algorithm. For compression to work, both the sending and receiving ends must use the same compression method. The sending end looks at the data, voice or image. It then codes it using a compression algorithm. The receiving end of the transmission decodes the transmission. For devices from multiple manufacturers to interoperate together, compression standards have been agreed upon for modems, digital television, video teleconferencing and other devices. The following are sample standards:

- *MNP 5:* Microcom Network Protocol compression protocol developed by Microcom for modems. Provides 2:1 compression.
- *V.42bis:* Data compression protocol for modems. Provides 4:1 compression.
- *H.320:* A family of standards for video adopted by the ITU, International Telecommunications Union. Quality is not as high as proprietary video compression algorithms. Most video codecs employ both proprietary and standard compression algorithms. The proprietary compression is used to transmit to another "like" video unit and the H.320 standard algorithm is used when conferencing with a different brand unit.

Digital Television—Sending Studio-quality Pictures with Compression

Compression squeezes video and analog signals into small enough units so that studio-quality television can be sent on standard digital television channels. The analog standard for television is set at 525 scan lines, or 525 lines of image. HDTV, high-definition television, will enable a TV screen to display 1125 scanned lines. A higher number of scan lines result in a clearer, studio-quality TV picture. Additional "lines" of image are seen as a denser, higher resolution of detailed images on the screen. This is done through computer manipulation of the video and audio portions of the television signal. Computerized compression takes out the redundancy and images in the picture that don't change. This reduces the signal that needs to be transmitted from 1.5 Gigabits to 19.3 megabits. However, the person seeing the TV image perceives the image to be almost as clear as the originating program. Because of the powerful compression and decompression tools, very little is lost to the viewer. The quality on digital television will be such that people watching television in their homes will perceive the quality to be like that of movies at theaters.

Multiplexing—Let's Share

Multiplexing combines traffic from multiple telephones or data devices into one stream so that many devices can share a telecommunications path. Like compression, multiplexing makes more efficient use of telephone lines. However, unlike compression, multiplexing does not alter the actual data sent. Multiplexing equipment is located in long distance companies, local telephone companies and at end-user premises. It is associated with both analog and digital services. Examples of multiplexing over digital facilities include T-1, fractional T-1, T-3, ISDN and ATM technologies.

The oldest multiplexing techniques were devised by AT&T for use with analog voice services. The goal was to make more efficient use of the most expensive portion of the public telephone network, the outside wires used to connect homes and telephone offices to each other. This analog technique was referred to as frequency division multiplexing. It allowed multiple voice and later data calls to share paths between central offices. Thus, AT&T did not need to construct a cable connection for each conversation. Rather, multiple conversations could share the same wire between telephone company central offices.

Digital multiplexing schemes also enable multiple pieces of voice and data to share one path. Digital multiplexing schemes operate at higher speeds and carry more traffic than analog multiplexing. For example, T-3 carries 672 conversations over one line at a speed of 45 megabits per second (see Figure 1.4). With both digital and analog multiplexing, a matching multiplexer is required at both the sending and receiving ends of the communications channel.

While T-3 is used for very large customers and for telephone company and Internet service provider networks, T-1 is the most common form of multiplexing for end-user organizations. T-1 is lower in both cost and capacity than T-3. T-1 allows 24 voice and/or

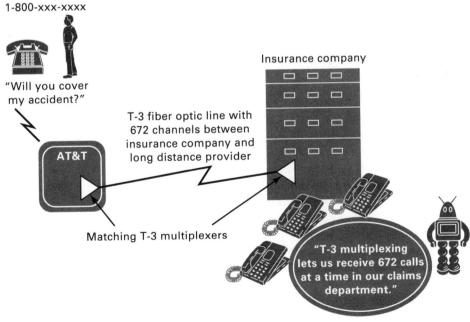

Figure 1.4 Multiplexers for sharing a telephone line.

data conversations to share one path. T-1 applications include linking organization sites together for voice calls, email, database access and links between end-users and telephone companies for discounted rates on telephone calls. Like T-3 services, matching multiplexers are required at both ends of a T-1 link.

Bauds, Bits, Bytes and Codes—Getting Down to Basics

Overview

Computers communicate using digital signals called bits. Bits are binary. They take two forms, on and off. Computers can "read" each others' communications when these bits are arranged in a standard, predefined series of on and off bits. All English language IBM and Mac computers use variations of the same type of codes. The main code, ASCII, is used when personal computers communicate over telephone lines. IBM minis and mainframes use a different code, EBCDIC.

People use the terms bits, baud rate and bytes interchangeably. Their meaning, however, differs significantly. The signaling speed on analog lines is the baud rate. The baud rate is measured differently than bits per second. Bits per second are the actual number of bits sent in a given time from point A to point B. It is the amount of information or data transmitted on the electrical waves in analog telephone lines.

Baud Rate vs. Bits per Second—Signal vs. Amount of Information Sent

A baud is one analog electrical signal or wave. One cycle of an analog wave equals one baud. A complete cycle starts at zero voltage, goes to the highest voltage and down to the lowest negative voltage and back to zero voltage. A 1200-baud line means that the analog wave completes 1200 cycles in one second. A 2400-baud line completes 2400 wave cycles in one second. The term baud rate refers only to analog electrical signals. It does not indicate the amount of information sent on these waves.

The public switched network runs at 2400 baud. If the public network could carry only 2400 bits in one second, data communications users would be severely hampered in retrieving and sending information over analog lines. To achieve greater capacity, modem manufacturers design modems capable of adding more than one bit to each analog wave or baud. Thus, a 9600 bit per second modem enables each analog wave to carry four bits of data per wave (9600 ÷ 2400 = 4). It is correct to state that the 9600 bit per second modem runs at 2400 bauds per second. A 28,800 bit per second modem puts twelve bits of data onto each electrical signal or wave. It still uses a 2400-baud line.

Baud rate refers to analog, not digital transmission services. Digital services do not use waves to carry information. Information is carried as on or off electrical signals in the case of copper wires, and on or off light pulses on fiber optic lines. On digital services,

Sending Attachments with Email

Email is the most widely used application on the Internet. However, email has format limitations. It only sends ASCII code. The limitation with ASCII is that it has just 128 characters. These characters do not include bold characters, images, tables or spreadsheet formats. This is a problem for people who want to conduct business or exchange complex documents.

For example, for my teaching at Northeastern, students send me their finals and I send consulting proposals and completed reports to clients and prospective clients. These files are usually in Microsoft® Word® or Microsoft Excel® format. Salespeople may send or receive presentations composed in the Microsoft PowerPoint® format. It is possible to exchange video, audio and JIF or JPEG image files.

To overcome ASCII limitations, mail protocols allow users to send attachments over communications lines. The mail protocol adds special bits to the beginning of the attachment which contains the word processing, spreadsheet or image file. These special bits tell the receiving computer when the attachment begins and ends and the type of encoding used, e.g., word processing program, spreadsheet, image, etc. The receiving computer then opens that particular program, i.e., spreadsheet, PowerPoint, JPEG or video, and decodes the attachment so that the recipient can read the document.

56,000 bit per second lines can carry 56,000 bits in one second. The speed is 56 Kbps, or 56 kilobits per second.

Codes—Adding Meaning to Bits

To enable computers to converse in a common "language," digital bits are arranged in codes such as ASCII for personal computers and EBCDIC for IBM mainframes and mini-computers. Codes allow computers to translate binary off and on bits into information. For example, simple email messages can be read by distant computers because they are both in ASCII. ASCII, American Standard Code for Information Interchange, is a seven-bit code used by PCs. ASCII code is limited to 128 characters. These characters include all of the upper- and lower-case letters of the alphabet, numbers and punctuation such as !, " and : (see Table 1.2).

Table 1.2 Examples of ASCII Code

Character	ASCII Representation
!	0100001
A	1000001
m	1101101

Because there are only 128 characters in ASCII, formatting such as bolding, under-lining, tabs and columns are not included in ASCII code. Specialized word processing and spreadsheet programs add their own code to ASCII to include formatting and specialized features. Thus, Microsoft Word documents, for example, need to be "translated" if they are to be "read" by a WordPerfect program. Each program uses a different arrangement of bits to, for example, format columns, tabs and footers. They each add proprietary formatting code to standard ASCII code. Sending documents between computers in ASCII allows them to be read by all PCs. However, specialized formatting such as tabs, tables, columns and bolding are not included in the transmission.

Bytes = Characters

Each character of computer-generated code is called a byte. A bit is only an on or off signal. The entire character is a byte. A one-page document might have 250 words with an average of five letters per word. This equates to 5×250, or 1250 bytes or characters. It would, however, contain 8,750 bits if each character were made up of seven bits. To sum-marize, a byte is a character made up of seven or eight bits. A bit is an on or off signal. Table 1.3 contains definitions of various network terms.

Table 1.3   LANs, MANs and WANs—What's the Difference?

Term	Definition
LAN Local Area Network	A group of data devices, such as computers, printers and scanners, that can communicate with each other within a limited geographic area such as a floor, department or building.
MAN Metropolitan Area Network	A group of data devices, such as LANs, that can communicate with each other within a city or a large campus area covering many city blocks.
WAN Wide Area Network	A group of data devices, such as LANs, that can communicate with each other from multiple cities.
Hub	The intelligent wiring center to which all devices, printers, scanners, PCs, etc., are connected within a segment of a LAN. Hubs enable LANs to be connected to twisted pair cabling instead of coaxial cable. Only one device at a time can transmit via a hub. Hubs provide a point for troubleshooting and relocating devices. Speed is usually 10 Mbps.
Switching Hub	Switching hubs allow multiple simultaneous transmissions on a LAN segment. Total speeds range from 10 Mbps to 100 Mbps (megabits per second).
Backbone	Wiring running from floor to floor in single buildings and from building to building within campuses. A backbone connects to hubs located in wiring closets on each floor.

Table 1.3 LANs, MANs and WANs—What's the Difference? *(Continued)*

Term	Definition
Bridge	Bridges connect multiple LANs together. They have limited intelligence and generally only connect a few LANs together. Bridges were in limited use as of the early 1990s when the price of routers dropped.
Router	Routers connect multiple LANs. They are more complex than bridges and can handle a greater number of protocols and LANs. Routers select the best available path over which to send data between LANs.
Server	A centrally located computer with common departmental or organizational files, such as personnel records, sales data, price lists, student information and medial records. The server connects to a hub. Access may be restricted.

The difference between LANs, MANs and WANs is the distance over which devices can communicate with others. As the name implies, a local area network is local in nature. It is owned by one organization and is located in a limited geographic area, usually a single building. In larger organizations, LANs can be linked together within a complex of buildings on a campus. Devices such as computers linked together within a city or metropolitan area are part of a metropolitan area network. Similarly, devices that are linked together between cities are part of a wide area network.

LANs—Local Area Networks

Examples of devices within LANs that can communicate are: shared printers, PCs, alarm devices, factory automation and quality control systems, shared databases, factory and retail scanners and security monitors (see Figure 1.5). A discrete LAN is typically located on the same floor or within the same department of an organization.

The growth of LANs grew out of the proliferation of PCs. Once people had PCs on their desktops, the next step was to connect these PCs together. LANs first appeared in 1980. The initial impetus for tying PCs together was for the purpose of sharing costly peripherals, such as high-speed printers. LANs are now the building blocks for connecting multiple locations together for the purpose of sending email and sharing databases with remote locations and telecommuters. These email and corporate information files are located in specialized computers called file servers. Access to file servers can be limited by password to only certain users

Devices on local area networks are all connected to the LAN. Each device on a local area network can communicate with every other device. The connections between devices may be any of the following: twisted pair, coaxial cable, fiber optics or wireless media. For the most part, devices are connected to a LAN by twisted pair cabling, similar but of a

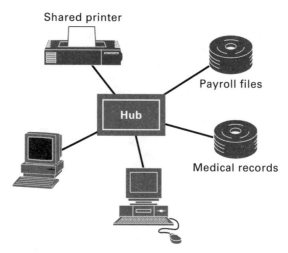

Figure 1.5   A local area network.

higher quality to that used to tie business telephones together. (Media options are covered in Chapter 2.)

When local area networks became popular in the 1980s, many individual departments purchased their own LANs independent of the central computer operations staff. As the need arose to tie these LANs together for email and file sharing, compatibility between LANs from different manufacturers became a problem. The TCP/IP suite of protocols became a popular choice for overcoming these incompatibilities. Devices called bridges and routers were also developed to send data between LANs.

Hubs

Hubs enable devices on local area networks to be linked together by twisted copper pair wire instead of the heavier, fatter coaxial cable typically used in the cable TV industry. When LANs were initially implemented, they were installed using coaxial cable to interconnect devices on the LAN. Coaxial cable is expensive to install and move. It is not unusual in large organizations for entire departments and individuals to move at least once a year. Use of coaxial cabling resulted in running out of space in dropped ceilings and conduit for the cable.

With a hub, instead of wiring devices to each other, each node or device is wired back to the hub. Using a hub changes the topology of a local area network. The hub creates a star design or topology. (Topology is "the view from above." In the case of hubs, a star.) Without a hub, each device in a LAN is wired to another device in a "bus" arrangement. In the bus topology, if one device is taken out of the line or bus, each device is affected. Conversely, by employing a hub, moving a device does not impact the other devices. A hub is kept in the wiring closet of each floor within a building, as shown in Figure 1.6.

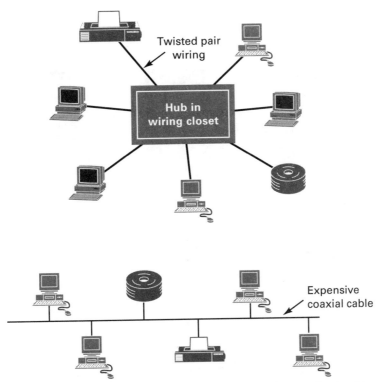

Figure 1.6 *Top*: LAN with a hub to link devices with twisted pair wiring. *Bottom*: LAN without a hub.

Bridges

Bridges became available in the 1980s as a way to connect a small number of LANs together. They were used most often in the mid-1980s. Bridges provide one common path over which multiple local area networks may be connected together (see Figure 1.7). For example, if an organization has two locations in different cities that need to exchange data, a bridge can be used. Bridges can connect two Ethernet LANs, or an IBM token ring network to an Ethernet LAN. In addition to connecting distant LANs to each other, bridges were used extensively in the mid-1980s to connect LANs in the same building or campus.

The advantage of bridges is that they are easy to configure. There are a limited number of choices in configuring a bridge. Each piece of data sent via a bridge takes the same path. This is also a disadvantage. Each piece of data not only takes the same path, it is also sent to each device on the network. The lack of routing and congestion control puts bridges at Layer 2 in the OSI model. Only the device to which the message is addressed takes the message off the network. This broadcast feature of bridges can choke the network with too many messages, slowing down the network for everyone. As LANs proliferated and router prices dropped, people turned to routers rather than bridges.

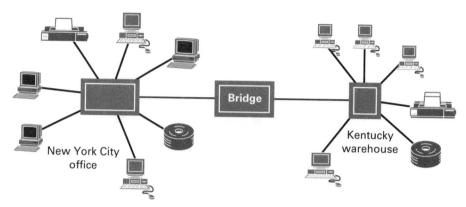

Figure 1.7   A bridge connecting two local area networks.

Routers

Routers are also used to connect multiple local area networks. These LAN connections are usually between LANs located in distant buildings on a campus or in different buildings in diverse cities. However, routers also connect multiple LANs within large campuses spread out across cities. Routers are more sophisticated and have additional capabilities not available in bridges. A major advantage of routers is their ability to forward differing protocols from varied departmental local area networks. It is important to note that routers do not translate application protocols. A UNIX computer cannot read a Microsoft Windows word processing document. The router merely allows differing LAN protocols to be transported via a corporate network infrastructure.

Router capabilities include:

- *Flow control:* If the path the data should take is congested, the router can hold the data until capacity is available on the path between the routers.
- *Path optimization:* The sending router selects the best available path. It checks routing tables contained within the router for this information.
- *Sequencing:* Routers send data in packets, or envelopes. These packets may arrive out of order at the end router. The receiving router knows by information in the packet the correct order and arranges the data accordingly.
- *Receipt acknowledgment:* The receiving router sends a message to the sending router letting it know that data was received correctly.

The above intelligence inherent in routers leads to two major disadvantages. In the first place, routers are complex to install and maintain. Every router in an organization's network must have up-to-date address tables. Each device on a local area network is called a node and has an address. For example, if a printer or PC is moved from one LAN to another, the router table must be updated or messages will not reach that device. To illustrate the complexity of managing routers, it is common to hear of consultants with full-time contracts to update router tables for organizations. Secondly, routers are slower than bridges. The need to look up tables within the router slows down the router's speed. The above functionality of congestion control, sequencing and receipt acknowledgment make routers network Layer 3 devices in the OSI model.

WANs—Wide Area Networks

The term WAN refers to connections between organizational locations over long distances via telephone lines. For example, a warehouse in Alabama connected to a sales office in Massachusetts by a telephone line is a WAN, or wide area network connection. In contrast to a local area network, a WAN is not contained within a limited geographical location. The variety of WAN connections available is complex. Selection of an appropriate WAN connection depends on the amount of traffic between locations, quality of service needed, and price and compatibility with the computer systems located within the organizations. WAN technologies and WAN vendors are reviewed in Chapters 6 and 7. These include: ISDN, T-1, T-3, ATM and frame relay. as well as wireless services.

MANs—Metropolitan Area Networks

Metropolitan area networks, MANs, are connections between local area networks which occur within a city or over a campus. Campus MANs are spread out over many blocks of a city. Examples of MAN networks are those of large hospitals and university complexes. For example, a hospital in downtown Boston keeps its x-rays and other records in a nearby section of the city. Instead of trucking records and x-rays between the two sites, the hospital leases high-capacity telephone lines to transmit records and images. The connections between these two sites are metropolitan area connections. These connections can be leased from a telephone company or constructed by the organization. They may be fiber optic, copper or microwave-based services. They may also include the same services mentioned for WANs, such as ISDN and T-1.

LAN and WAN Congestion

New, High-bandwidth Applications

Original LAN designs lent themselves to "bursty" traffic. Bursty traffic includes email and text messages. Bursty traffic is not a steady stream of data. With typical LAN protocols, such as Ethernet and token ring, only one message at a time can be carried on a LAN that has a speed of ten megabits. New applications are causing delays and congestion on LANs. Applications adding high-traffic volumes to LANs are: desktop video conferencing, computer aided design, computer aided manufacturing and graphics downloaded from the Internet.

Not only are these applications adding traffic to LANs, but the traffic is no longer the short, bursty type of traffic. Bursty traffic sends a group of messages and then has a pause. This pause gives other devices that share the network a chance to transmit data. Video, however, is an application that requires constant use of the network. People participating in a conference don't want a blank screen while someone else on the LAN accesses the Internet. Video requires constant use of the network during the video conference.

More Powerful PCs

In addition to applications which require large amounts of data to be transmitted over organizations' LANs, the capability of PCs impacts LAN requirements. In the 1980s when LANs were first implemented, people had 286 computers on their desks with small amounts of memory and hard disks. In recent years, staff have 486 and Pentium computers with 16 megabits of memory and Gigabit-sized hard drives. These powerful PCs have multimedia capability. This allows them to participate in desktop video conferences, download large files from the Internet and share large spreadsheet files. All of this traffic is carried over the LAN.

Sharing the LAN

Router-based and hub-based campus networks and local area networks are shared media networks. Everyone has a turn to send and receive data, but sharing is required. Only one message at a time can be carried. The speed on these networks is high, 10 megabits. But, the assumption is that messages will be bursty, allowing other transmissions to send without causing large delays. When LANs were first implemented, in addition to assumptions regarding burstiness, it was assumed that applications would not require immediate response. For example, email transmissions could be delayed without impacting customer satisfaction. This is not true for newer applications such as Internet access. People do not want delays when downloading information from the World Wide Web.

Congestion within LANs, LAN-to-LAN and LAN-to-WAN

Congestion on networks occurs both within a local area network, between LANs in a building or campus and between a LAN and a WAN. New technologies are emerging which provide greater capacity in these areas.

Emerging services for LAN traffic (All Require Hub Upgrades)

- *Fast Ethernet:* Fast Ethernet is a shared protocol. However, it has a speed of 100 megabits—ten times the speed of standard Ethernet, the most prevalent LAN protocol. Standard two pair wiring is used. New cards are required in each PC to access the LAN.

- *Ten-megabit Switched Ethernet:* Switched Ethernet is a non-sharing service. Devices with high transmission needs are given their own dedicated paths within a LAN. Standard wiring, bridges and routers can be used. This frees up high bandwidth users from "hogging" LANs.

- *100VG/AnyLAN:* Like switched Ethernet, gives heavy users dedicated paths to each other on LANs. Required wiring is somewhat non-standard—four pairs of copper wire. Many locations have only three pairs of wire to each device. Promulgated by Hewlett-Packard and more complex than Switched Ethernet.

Emerging Services for LAN-to-LAN Backbone and LAN-to-WAN Traffic

- *Gigabit Ethernet:* Works with existing LAN protocols. Fiber optics required because of its high speed of 1000 megabits. This is a "shared" data transmission.
- *IP switching:* Introduced by Ipsilon Networks Inc. in December of 1996, IP switching works only with TCP/IP networking protocols. Faster than routers because IP switching does not have to "sniff" messages to know which protocol they are associated with. Cheaper than routers also.
- *Tag switching:* Supported by Cisco. A proprietary protocol to increase the speed of connections between LANs.
- *IP/IPX switching:* Supported by Bay Networks. Takes WAN traffic from routers and switches it to local area network segments. Speeds up the transfer from the WAN to individual devices.

CHAPTER 2

Telephone Systems
and Cabling

T his chapter covers on-site telephone systems and peripheral devices for telephone systems, including voice mail and call center services. Also covered are the copper and fiber cables that connect telephones to telephone system controllers and computers to other data devices. The market for on-site telephone systems such as PBXs, key systems, voice mail and call center services is highly competitive. However, end-users are often confused about the type of system to buy and how all of the features work. They are bombarded by salespeople who offer deep discounts but few explanations of what the features mean. For example, how user-friendly are the telephones, how does the voice mail sound to callers, how much capacity is available in the system?

The first issue people purchasing telephone systems are faced with is understanding the difference between private branch exchanges, PBXs, Centrex and key systems. All of these systems provide connections between staff at organizational sites and the outside world. They additionally are the means for on-site personnel to call other personnel at their site without paying telephone company usage fees. The difference between PBXs and Centrex is in the location and ownership of the equipment that routes calls. Private branch exchanges are located on customers' premises. Centrex, which stands for central exchange, is located at the telephone company. It is part of the central exchange or central office. In rare cases with large Centrex systems with over 2000 telephones, Centrex is located at the customer's premise. In the vast majority of Centrex installations, however, the equipment is at the central office. Nevertheless, in either case, the customer leases Centrex service and does not own the equipment used to route calls. This equipment is owned by the telephone company that provides the Centrex system.

Key systems function much the same as private branch exchanges, PBXs. They are located at customers' premises. Often, key systems are smaller than PBXs. Key systems formerly had fewer features and less functionality than PBXs. This is changing. They now have sophisticated voice mail, services for call centers and telephones equipped with features such as speed dial and redial. Key systems are sold most often to organizations with less than 100 telephones at a site. Systems with between 50 and 125 telephones fit into both key system and PBX configurations depending on the features in the particular telephone system. For example, some key systems have less sophisticated voice mail or call center services. The lines of functionality between key systems and PBXs are blurring as larger key systems with more features are being developed.

Peripheral devices, such as those used in call centers, are often high-priced and provide large margins to vendors. These systems are called automatic call distribution systems, ACDs. ACDs route incoming calls to agents based on criteria such as the agent that has been idle the longest. If an agent is not available to take a call, the automatic call distributor holds the call in a queue and the caller hears a message such as, "Please hold for our next available agent." ACDs are sold as part of key systems, Centrex and PBXs. They are also sold to the most sophisticated call centers as stand-alone systems. Organizations pay high prices, up to $2,000 per telephone, for automatic call distributors because of their value in managing staffing and telephone lines, the two most expensive portions of a call center. They do this by providing, in addition to call queuing, sophisticated reports on agents and telephone lines. Managers have information on the number of agents and telephone lines needed to handle particular levels of calls.

The Second College Edition of *The American Heritage Dictionary* defines medium as "An intervening substance through which something is transmitted or carried on..." Telecommunications media carry voice, data, video and images. Examples of media are: fiber optic cabling, twisted pair (copper), air waves (e.g., microwave and cellular telephone service) and coaxial cabling. The quality and type of media used to carry transmissions have an impact on the speed that the data can travel, the amount of errors and reliability of services. This chapter covers fiber optic and twisted pair media and touches upon wireless services for local area networks inside buildings and in conjunction with telephone systems. Chapter 9 explores wireless for cellular and satellite services.

Telephone Systems: From Stand-Alone to Connected Telephones

When the telephone was first invented in 1876, each person's telephone line was wired directly to another individual user. By 1877, a switchboard was installed in Boston so that each telephone would not have to be wired to each other telephone. Rather, each telephone was wired from its building to the central "switchboard." When an individual wished to call someone, he or she picked up the telephone handset and asked the operator to connect his or her call to another specific individual. The operator knew all of the town's business. In 1891, Almon Strowger patented a central office switch. Strowger's motivation was privacy. He felt that operators were listening in on his telephone conversations.

What Is a PBX?

The central office switch is the precursor to on-site PBX telephone systems private branch exchange. Where a central office switch is located centrally and routes calls in the public network between users, PBXs are private. They are located within a specific location. PBXs route calls:

- Between people located within an organization.
- From individual users within an organization to people outside of the system.
- From callers outside the location to on-site individuals.

Just as a central office switch eliminates the need to wire each telephone to each other telephone, a PBX eliminates the need to wire each on-site individual telephone to each telephone in the company. It also eliminates the requirement to wire each telephone to the central office. In essence, with a PBX, each employee does not have to pay for his or her own telephone line from the local telephone company back to his or her office. Nor are there charges for calls between people in the same office. Depending on the volume of calls generated by the staff, each telephone path between the PBX and central office can be shared by eight to ten users. A PBX with 100 users might share 12 paths, called trunks, for outside calls.

Connecting Telephone Lines to PBXs

The local telephone company brings telephone lines into a building to an interface to which the outside lines are wired. This interface is called a jack or a punch down block because each outside line is punched down to the connecting block. Jacks that hold one line are called rj11c jacks. These are the jacks found in most homes. The most common interface to which local telephone companies wire multiple outside lines in businesses is the rj21x. This interface holds 25 lines. The importance of the rj21x jack is that it is a common point from which telephone lines and trunks can be tested. For instance, if there is a question on a repair problem as to where the problem lies, the telephone company can test its line to the jack and the PBX vendor can test service up to the interface. The rj21x jack is the demarcation point between the telephone company line and the inside wiring (see Figure 2.1).

PBX Telephones

When PBXs were first installed, everyone had the same "vanilla" type of rotary telephone called a 500 set. Touch-tone, single-line telephones are still known as 2500 sets. AT&T set the standard for telephones and AT&T telephones were the only products available in AT&T-controlled areas. AT&T eventually found a way to provide users with electromechanical telephones capable of holding up to nine lines on one telephone. A hold button enabled users to put callers on hold to answer calls on the same telephone.

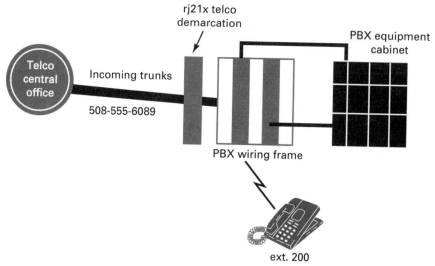

Figure 2.1 A PBX connection to the central office.

By the 1970s, other manufacturers such as Rolm offered competing PBXs. This was the start of the interconnect industry which supplies customer premise equipment, CPE. On-site telephone systems have become more complex over the years. More features have often been synonymous with more complexity for the end-user. Many people are nostalgic for the days when using a phone was as easy as pushing a button on a telephone and dialing a call. Everyone had the same type of "plain vanilla" telephone and they all worked the same way.

PBX features have grown to such an extent that what was once simple is now complex. A major challenge in the last decade has been to simplify the use of advanced features associated with telephone systems.

Centrex

Whereas PBX stands for private branch exchange, Centrex is derived from the words "central exchange." In contrast to PBXs, where the switching equipment is located on organizations' sites, Centrex switching equipment is part of the central office. Thus, the Centrex nomenclature central exchange.

The first fully automated Centrex system was installed in 1965 at Prudential Life Insurance Company in Newark, New Jersey. The original motivation for Centrex is much the same as the desire for Centrex and also voice mail automated services today. Organizations wished to save money on operators, administration and space. Centrex provides four ways of fulfilling these savings:

• *No requirement for on-site switching equipment:* Main Centrex switching equipment is in the telephone company's central office.

User-friendly Technology

Creating user-friendly telecommunications devices is a major challenge and an important factor in user acceptance of new technology. One of the reasons for Rolm's popularity, in addition to smart marketing, in the early days of the interconnect industry, was that its telephones were good-looking and easy-to-use. Features such as conference calling, speed dialing and transfer were available at the touch of a button on the telephone. Unfortunately, these telephones themselves were so expensive that most organizations installed only about 20% of these feature phones and the rest were "vanilla" type 2500 sets.

A frequent complaint heard about telephone systems is that they are hard to use and that people only use a few of the myriad features. A recent improvement is feature-prompting done via the liquid crystal display of high-end telephones. For example, when a telephone user presses the transfer button to transfer a call, the LCD displays a prompt such as, "dial the number to which the call will be transferred." Creating easy-to-use technology goes a long way to ensuring its acceptance and adoption by end-users. As manufacturing costs decrease and prices are lowered, organizations are purchasing more of these high-end telephones when they buy new systems.

- *Direct-inward dialing:* Centrex incoming calls can go directly to telephone users without being answered first by the company operator.
- *Direct-outward dialing:* On-site telephone users can dial outgoing calls without having an operator place the calls for them.
- *Automatic identification of dialed calls:* The telephone company bill identifies the telephone extension from which each outgoing toll call was made.

Centrex service appealed to customers who thought that automating the dialing of incoming and outgoing calls would save them money on operator costs. An operator would not be required to answer all incoming calls and to complete all outgoing calls. Moreover, accountability on toll costs would be preserved by telephone company billing which indicated who made each call. In addition, organizations did not have to dedicate large rooms to the old electromechanical switches used to handle corporate calling. This saved real estate costs associated with supporting on-site PBXs.

AT&T had additional motivations for promoting Centrex which, according to a 1986 DataPro report, provided service to 70% of all businesses with over 1,000 lines until 1982. These motivations were:

- Stimulation of toll calling.
- Less use of long distance circuits for call setup.
- A reduction in the required number of toll operators to manually place calls.

Automating the ability to dial calls without operator intervention made toll calling more accessible and stimulated usage. Automating the process also provided more efficient use of expensive telephone lines for non-billable call setup. When an AT&T operator places a calls for a user, non-billable time is used to provide the operator the number and have the operator manually dial the call for the end-user.

Centrex Sales Channels

Prior to the 1984 divestiture, AT&T owned both the local telephone companies and the long distance and customer service equipment channels of AT&T, and the local telephone companies such as New York Telephone. Centrex was the main telephone system sold by AT&T and its 22 local telephone companies such as Michigan Bell and Illinois Bell. Beginning in the late 1970s, AT&T began changing its strategy from pushing Centrex to favoring rental of stored program-controlled, on-site telephone systems. It felt that Centrex, with its requirements for a separate pair of wires to each telephone from the central office, wasted the most expensive portion of the telephone company operation, outside cabling. A 2,000-line Centrex system required 2,000 cable runs from the central office to the customer premise. In contrast, a 2,000-line PBX required approximately 200 cable pairs from the central office. (With a PBX, telephones access a shared pooled group of outside lines.) The remaining 1,800 cable pairs could be used by AT&T for other customers.

After divestiture, this strategy changed because the local telephone companies were no longer allowed to furnish on-site telephone systems. Not only did the telephone companies decide to promote Centrex heavily following the 1984 divestiture, they changed the target group to whom Centrex was promoted. Prior to divestiture, Centrex was aimed primarily at large, over 2,000 line organizations such as:

- Hospitals.
- Cities and towns.
- Universities.
- Large businesses located in campus environments.

Following divestiture, the local phone companies broadened their marketing of Centrex to the under 100 line market. The under 100 line market is the fastest growing; it has the largest number of sites; and thus the largest portion of the on-site telephone equipment market. Marketing campaigns stressed the following advantages for Centrex:

- Redundancy built into the central office switches.
- Space reduction requirements due to Centrex being located at the central office.
- Ease of administration—the phone company would take care of everything.
- Ease of growth that could be handled by growth in the central office.
- Compatibility with new network technologies such as ISDN.
- Fewer power and cabling expenses because of no on-site switching equipment.
- Vendor support—the telephone company would never go out of business.

While this strategy appealed to many small customers without the technical expertise to select and maintain an in-house telephone system, the local telephone companies did not have the marketing expertise to follow through effectively on this strategy. They turned to sales agents as a channel through which to sell both Centrex and all local and local toll calling services.

Sales agents generally also sold customer premise equipment such as PBXs, key systems, voice mail, cabling installation and equipment maintenance services. However, growth in customer premise telephone systems in the late 1980s was slow and margins were slim. Moreover, many organizations kept their phone systems for ten years. A sales agent who retained an organization as a customer received a monthly fee from the local phone company plus extra commissions for installation of services such as T-1 and data communications lines. In return, the sales agents placed all of the end-users' repair, installation and change orders with the telephone company.

Sales agents aggressively marketed telephone company services. They frequently offered to analyze a customer's telephone bill at no charge. There are two strategies they frequently used to sell Centrex. First, they offered Centrex to customers with existing telephone systems as a way to save money on telephone lines. (In these instances, Centrex with a three- to seven-year commitment may have been cheaper than plain old telephone lines.) They also promoted Centrex as an option for a new telephone system. On new Centrex sales, they sold peripheral equipment such as voice mail and upgraded, feature-rich telephones. These feature rich, often proprietary telephones may have required on-site key service units which worked in conjunction with the Centrex service (see below). In fact, many customers ended up with both Centrex service and on-site telephone systems.

Centrex Telephone Sets

Three choices are:

- Standard 2500 sets, the same type available for homes.
- Central office powered phones with features provided by the central office.
- Proprietary telephones with features provided by on-site key service units.

The limitation with standard 2500 "vanilla" type Centrex telephones is that they can each hold only one or two lines. While standard 2500 telephones are available with features such as speed dial, transfer and conference buttons, heavy telephone users such as sales and customer service staff often want to put one caller on hold to make or answer a second call. If a user has two lines, he or she must pay the monthly fee for two Centrex lines. In addition, these telephones lack the coverage capabilities of proprietary telephones. Coverage means the ability for departmental secretaries to have buttons on their telephones indicating whether people in the department are on their telephones or not.

Two ways to obtain call coverage and multi-line capabilities is by purchasing or leasing proprietary telephones. These telephones can work off the intelligence in the central office or the intelligence of an on-site key system processor. Multi-line telephones that work off central office intelligence are either ISDN telephones or Nortel's P phones (the "P" stands for proprietary). ISDN telephones require a monthly ISDN fee plus monthly fees for the features on the telephone, which is an expensive proposition.

Many manufacturers make specialized telephone systems for the express purpose of working with Centrex telephone lines. Key providers of these telephones are: Comdial, Nortel, Lucent, Tone Commander and Siemens. These systems have Centrex features programmed such that, for example, when a telephone user transfers a call, the transfer signal is sent to the central office. This tells the central office to "free up" the transferring person's telephone by correctly transferring the call to another user. If the transfer is not done via the central office, the telephone call continues to tie up the telephone user's set that transferred the call.

On-site telephone systems provide the call coverage capability requested by heavy telephone users. The central processing unit of the telephone system can be programmed to ring and send a call to multiple instruments. A departmental secretary can, for example, have a button associated with each extension of each manager in the department. The central processor of the on-site telephone system can send conversations onto the Centrex lines such that multiple calls can be handled by a single instrument within a department. A single Centrex line is terminated onto equipment on a telephone system called a line card. Each of the from four to eight ports on the line card is associated with one telephone. However, the same extension can appear on multiple telephones.

An ISDN telephone, which gets its features from the central office, has even more functionality. The same telephone wires that carry each Centrex line from the central office can carry multiple conversations. The downside is that the telephone company charges extra monthly fees for this functionality. Moreover, ISDN is not available from all central offices. For the offices where ISDN is available, there is a distance limitation that prohibits customers far from the central office from having ISDN. The end result is that customers that have Centrex often also end up needing customer premise equipment as well as the Centrex service.

Key Systems

Like PBXs, key systems are on-site telephone systems that route calls between people within organizations and route calls to and from staff from the public network. While there are some technical differences in the way they handle calls, new key systems have all of the features and most of the functionality of private branch exchanges. Key systems generally serve the under 70 user per site market.

The major difference between key systems and PBXs is the connection between the central office and the key system. Key systems are loop start and PBXs are ground start. What this means is that with a ground start PBX, a trunk or outside path is seized or grounded by the PBX or central office before a call is sent between the two locations. With a loop start key system, if a path is available, the call is sent either to the key system from the central office or to the public network from the key system. Analog home phones are also loop start, which is why a person can pick up the handset to make a call and find that someone calling them is already on the phone even though the telephone has not rung.

The PBX provides the dial tone to the user. On a key system, dial tone is derived from the central office. A person making an outside call on a key system does not have to dial an access code such as the number nine. Pressing an outside line button on a key system telephone signals the central office that the end-user wishes to make or receive a telephone call. On a PBX, users must first dial an access code, usually nine, to make an outside call. The PBX responds to a lifted handset by sending a dial tone to the end-user and then requesting that a trunk to the central office be "grounded" or seized to make a telephone call. This is the reason why, on key systems, end-users use an outside line button to make or receive outside calls and an intercom button for internal calls and transferred calls.

Other than these differences, to an end-user, new key systems provide all of the functionality of a PBX. In fact, many larger key systems are "hybrid" systems. They can be installed as either key systems with outside lines or PBXs with trunk connections to the central office.

In-Building Wireless PBX and Key System Telephones—On-site Mobility

Anyone who has waited for a specific telephone call knows that as soon as you step away from the telephone for a coffee break or to take part in a meeting, the telephone call you have been waiting for arrives. Staff such as nurses, warehouse employees and technicians spend more time away from their desks than at their desks. Wireless telephones enable workers to be reached (and interrupted!) at all times.

Specialized wireless telephones associated with PBXs and key systems are high-profit margin peripherals. These telephones operate at higher frequencies than home telephones and have specialized features associated with particular PBXs and key telephone systems. These features include:

- Hold buttons.
- Some of these telephones function both inside and outside of the campus. The telephone senses when it is out of the range of the PBX and calls are then switched via a cellular telephone network.
- Speed dial buttons that allow abbreviated dialing of frequently called numbers.
- Liquid crystal displays to show the name of the person calling.
- Message waiting lights to indicate new voice mail messages.

The use of wireless telephones inside buildings requires special base stations with antennas located on every floor (refer to Figure 2.2). There are generally also special outside base stations with antennas for nearby outdoor areas between buildings on a campus. The base stations need to be wired with twisted pair to specialized circuit packs within the telephone system cabinet. On-site wireless telephone systems use a cellular digital switching technology similar to that described in Chapter 9 for PCS. Calls are automatically transferred between base stations when a user walks out of the range of a particular antenna.

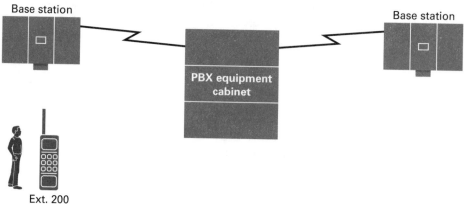

Figure 2.2   In-building wireless telephones.

Direct-inward Dialing—Bypassing the Operator for Incoming Calls

One of the most difficult facets of managing telephone systems is the ability to appropriately staff consoles to smoothly handle incoming calls. Frequently, organizations do not have information on the number of calls that hang up before they are answered or hear 15 or more rings before their call is answered. Many people managing telephone systems hear caller complaints that the company's telephone system must be broken because their calls are not being answered. In reality, often the consoles are so backed up with calls during peak periods that there are delays in answering.

One way to solve the problem of answering incoming calls is to install, from the local telephone company, a feature called direct-inward dialing (DID). This is also a key feature of Centrex service. Direct-inward dialing routes calls straight from outside lines to a telephone of a PBX or key system without operator intervention. Prior to the late 1980s, DID, direct-inward dialing, was priced so high from the local telephone companies, that only large organizations purchased it. Pricing has been lowered considerably and direct-inward dialing is now used by medium-sized key systems and PBXs. One small organization with only 25 telephone uses it to easily identify calls to the three companies it owns. Each of the three companies uses a separate telephone number within the DID series of 100 telephone numbers. The central office passes the dialed number to the key system so that the operator can answer the call in the name of the company called.

As Figure 2.3 illustrates, each telephone number does not use a separate pair of wires. Rather, organizations purchase groups of 100 "software" telephone numbers. They also order in the neighborhood of one wired path per eight to ten telephone extensions. The central office looks at the number dialed on the incoming call and identifies it as belonging to a particular organization. The central office then passes the last three or four digits of the dialed number to the organization's key system or PBX. The on-site telephone

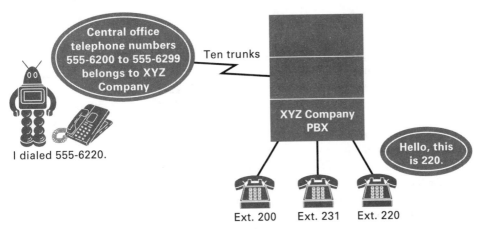

Figure 2.3 Direct-inward dialing from the central office to each telephone in a building.

system reads the digits and passes the call directly to the correct internal telephone. A small company with 25 telephones feels that having direct-inward dialing gives it the image of being a larger company. Calls go directly to the correct person and if they're not answered, they forward to voice mail. This is an example of technology coming down in price and being affordable to smaller organizations. (It's another reason why North America is running out of telephone numbers. Organizations are grabbing groups of 100 numbers for direct-inward dialing.)

Add-on Sales to Key Systems, PBXs and Centrex Systems

Sales of basic on-site telephone systems are extremely competitive and margins are slim. At the time of purchase, discounts of 40% to 50% are common. Vendors of telephone systems, customer premise equipment and CPE systems derive much of their profits in the following areas:

- *Maintenance* contracts on telephone systems.
- *Adds, moves and changes* of telephone equipment.
- *Upgrades* to the hardware and software of existing telephone systems.
- *Feature telephones* with built-in speakerphones, feature buttons and liquid crystal displays.
- *Wireless telephones* made for PBX and key systems.
- *Sale of peripheral devices* such as voice mail, automatic call distributors and call accounting systems.

Call Accounting—Tracking Who Made Calls and Usage on Telephone Lines

Call accounting systems track each telephone call made by individual users. They provide accountability on the use of the telephone and a way to charge back telephone costs to particular departments within an organization. They also indicate the amount of traffic on each telephone line so that organizations can determine when there are too many or too few outside telephone lines. A large amount of traffic during weekends or nights may also indicate fraudulent use of the telephone system. Hackers may be using the system to make free long distance calls. Fraudulent use of telephone lines for drug deals and the sale of international calling to people without telephones is not uncommon.

Call accounting, also called station message detail recording or SMDR, is a software package that generally sits on a PC. The computer collects call information through a connection to a port on the telephone system. An alternative to on-site call accounting systems is the use of service bureaus to collect calling statistics. Large organizations often collect this information by hiring an outside service bureau to collect the call data remotely via a port on the telephone system.

Voice Mail—A Way to Take Messages

Gordon Matthews, the founder of VMX, received the first patent on voice store and forward. The first VMX voice mail system was installed in 1980 at 3M. Other early vendors of voice mail were IBM through its Rolm division in 1982 and Wang Laboratories Inc. of Lowell, Massachusetts. Octel Communications Corp., founded in 1982, is a current leader in voice messaging and purchased VMX. Octel Communications Corp., through its purchase of VMX, owns the Matthews patent portfolio on basic voice store and forward. (Octel was purchased by Lucent Technologies in 1997.) All major manufacturers of proprietary and PC-based voice mail system pay licensing fees to Octel.

In the early to mid-1980s voice mail systems worked on proprietary computer platforms and used various techniques for connecting voice mail to PBX and Centrex systems. As the leading PBX vendors saw the success of voice mail, they jumped into the market. Vendors such as AT&T and Nortel sell their own proprietary voice mail and are popular with organizations that want the same vendor to support both their telephone and voice mail systems. In addition to customer-based voice mail, companies such as Boston Technology, Centigram and Octel Communications Corp. sell voice mail to telephone companies, cellular telephone providers and service bureaus who resell voice mail boxes to the public.

Customer premise voice mail changed from mainly proprietary systems, based on computers such as DEC, to PC-based systems. As PCs became more powerful, with larger hard drives and bigger amounts of random access memory, PC-based voice mail systems gained market share. Active Voice, Applied Voice Technology, AVT, Brooktrout Technology and Voysys Corporation all sell PC-based systems. Instead of developing their own PC-based voice mail major vendors either bought out companies who had already developed PC-based systems, or in the case of PBX and key system manufacturers Lucent

Technologies, formerly AT&T, Nortel, Toshiba and Comdial, put their names on third-party, PC-based voice mail systems via OEM arrangements. Lucent and Nortel use PC-based voice mail for their key systems. Octel acquired their PC-based voice mail by purchasing Compass Technology Inc.

Many companies, desiring the added productivity provided by voice mail, installed voice mail on older key systems. This was often a disaster. The older telephone systems were not capable of sending or receiving the voice mail signals necessary to:

- Turn message waiting indicators on and off.
- Allow users dumped into voice mail to access company operators.
- Send unanswered calls directly to the correct greeting. For example, a company operator might transfer a caller to extension 200. Without the signaling available between newer key systems and voice mail, if extension 200 did not answer, the caller would be answered by voice mail and hear, "Dial the voice mail number of the person you wish to leave a message for." With correct signaling, the caller would be answered, "This is Harry, I can't answer your call, etc."

The availability of lower-priced PC-based voice mail is a major factor in the decision to purchase new key systems. The lower price of PC-based voice mail makes it affordable to smaller organizations. Once voice mail prices dropped because of the lower cost of PC-based systems, small companies recognized the benefits derived from voice mail and purchased new key systems capable of working properly with voice mail.

Automated Attendants—Using Machines to Route Calls

Automated attendants are used as adjuncts to company operators. For example, an automated attendant is programmed to answer certain calls to particular telephone lines, e.g., "Thank you for calling ABC Company. If you know your party's extension, you may dial it now. For sales, press 1. For customer service, press 2." When first introduced in 1984, automated attendants were not part of voice mail systems. They consisted of separate hardware and software that connected to either the trunks or telephone lines of telephone systems and were programmed to answer all or some of an organization's outside calls.

The first automated attendant, manufactured by Dytel, was installed in 1984. At that time, many organizations had problems consistently answering calls within a reasonable number of rings. They found, for example, that in slow times such as 8 AM to 10 AM, operators had too much idle time. However, during the busy hours, 10 AM to 11:30 AM and in the early afternoon, company operators could not keep up with the call volume. Even during the busy hours, call volume was erratic with many peaks and valleys. Moreover, at that time, direct-inward dialing, DID, was still priced high by the local telephone companies. To solve these problems, many organizations used automated attendants to answer a special group of telephone numbers which they publicized to traveling employees such as salespeople and to vendors and spouses of employees. Thus, the calling volumes directed to live operators was lowered and first-time callers and customers had their calls answered more quickly.

Because they saw an immediate payoff in using automated attendants to reduce the workload and number of live operators, small and medium-sized organizations jumped on the automated attendant concept. Many of these smaller organizations bought automated attendants before they purchased voice mail. They saw the payback immediately in reduced labor expense. As the industry matured, automated attendant manufacturers were bought out by voice mail vendors and current systems have both voice mail and automated attendant functionality. Automated attendant is now a software feature of voice mail systems.

Voice Mail Components

Figure 2.4 illustrates a combination of the following:

1. *Central processing unit (CPU):* The CPU is responsible for the overall operation of the unit. It executes the application software and operating software that is located in the CPU.
2. *Codecs:* These devices convert analog voice to digital signals and digital signals back to analog. Most systems compress voice and take the pauses out of conversations to more economically store voice mail messages on the hard drive.
3. *Software:* The software distinguishes one system's features from another system's, e.g., the ability to automatically hear the time a message was left rather than having to dial a "7" to hear the time the message was left.
4. *I/O cards:* These printed circuit boards provide the connections between the telephone system and voice mail system. There are usually four ports per board. Each port enables one person to leave or pick up a voice mail message. I/O ports are also used for the receipt and transmission of facsimile messages on systems with voice mail as well as fax mail.
5. *Other system components:* Serial ports, scanners, high-speed buses, power supplies and tape and disc drives for system backups.

Using New Technology Without Offending People

When voice mail and automated attendants were first used in the 1980s, they created deep feelings and animosity among callers into these systems. Many people in the industry felt that the technical issues were easier than the "etiquette and procedural" issues. For example, many automated systems had long menus with a myriad of choices (some still do). In addition, finding a live body with whom to speak in an emergency was often impossible.

Antagonism to voice mail got so bad that in 1992, Octel Communications Corp. created a Voice Mail Education Committee with representatives from major voice mail manufacturers and service bureaus. They jointly set about contacting the media and doing studies on the best way to implement and use voice mail and automated attendants without offending customers. Octel went so far as to send out "I Love Voice Mail" pins to customers and distributors. Pacific Bell, as a result of the committee, published a *Tips and Etiquette* booklet and a *10 Easy Steps To "Caller Friendly" Voice Mail* card.

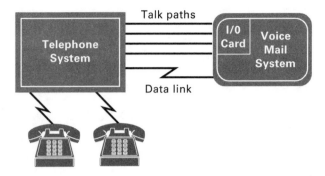

Data Link:

Telephone Number Dialed
Caller ID on internal calls
Messaging Waiting Notification On/Off
No Answer/Busy Notification On Called Party

Figure 2.4 Voice mail connections to a telephone system with signals between the voice mail and phone system. In small telephone systems, the voice mail system may be located within a shelf of the telephone system.

For organizations and residential customers who don't want to maintain on-site systems, voice mail services are available from service bureaus, telephone companies and cellular telephone providers. Boston Technology, Octel Communications Corp., Centigram Communications Corporation and Unisys all sell voice mail to telephone companies and service bureau-based voice mail. Cellular-based voice mail is particularly popular in developing countries without cabling infrastructures. The use of cellular central offices enables countries to rapidly deploy telephone service without laying miles of cable. Along with the cellular central office, international telephone companies are buying voice mail that works in conjunction with the cellular mobile switching office.

Systems Capable of Storing Voice Mail, Fax Mail and Email Messages

As people are becoming inundated with voice mail, facsimile and electronic mail messages, companies are looking at ways to consolidate messaging onto one platform. A feature of some newly installed voice mail systems is the ability to be notified of fax, email and voice mail messages on the same PC (see Figure 2.5). Thus, users don't have to access voice mail from their telephones, email from computers and facsimile messages from stand-alone fax machines. These voice mail systems store incoming facsimile messages on the voice mail system's hard drive. Voice mail users then issue commands from either their computers or their telephones, directing the voice mail system to print the faxes on particular printers. For example, users that travel can have fax messages printed at their hotels. The PC notifies users of multiple types of messages, but email and voice mail

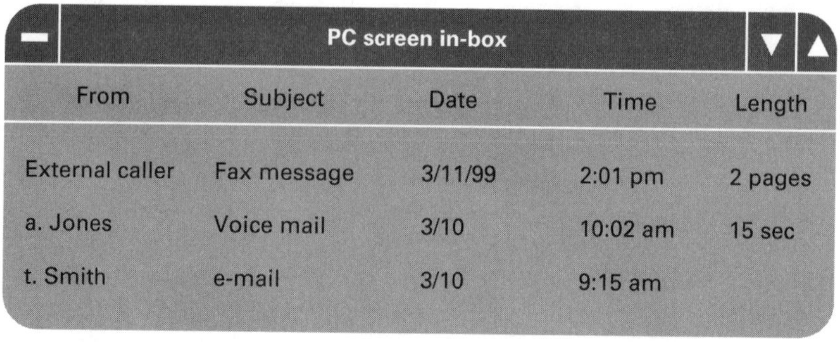

From	Subject	Date	Time	Length
External caller	Fax message	3/11/99	2:01 pm	2 pages
a. Jones	Voice mail	3/10	10:02 am	15 sec
t. Smith	e-mail	3/10	9:15 am	

PC screen in-box

Figure 2.5   Voice mail, fax mail and email notification from a PC screen.

may still be stored on different devices. Managers may still have to administer separate email and voice mail systems.

Further integration in messaging is planned to enable a single LAN server to hold directories and store fax mail, email and voice mail. Thus, the computer information system staffs will only be required to maintain one set of services for all types of messages. These messages will be accessible from multimedia PCs and from telephones. For example, someone could access an email message from his or her car telephone and hear a synthesized "voice" read the message to them. (The systems will be able to convert ASCII text to synthesized voice for voice playback of messages.) A major constraint in acceptance of these systems is the amount of bandwidth required to carry voice and facsimile messages on LANs.

To reduce the bandwidth, (i.e., capacity) required to transmit voice over LANs, voice mail vendors compress voice to make it smaller, so that it requires less bandwidth and hard drive space on the local network. Octel states that the average five minutes of voice storage needed by an average user of voice mail only requires 1.2 megabytes of storage. Furthermore, messages are "streamed" in segments, rather than as one large file, when they are played back, so only .4% of a 10 Mbit Ethernet network capacity is used to play a message and 10% to play back 24 voice/fax messages. This is another application where local area network capacity is an issue.

ACDs—Specialized Equipment to Handle Large Volumes of Calls

As the cost of making on-site sales calls escalates, more organizations are looking at ways to cut their selling costs. They also want to lower their cost of providing customer service to existing customers. In particular, they want to enable their employees to handle the maximum amount of customer calls. Employees are the largest expense in sales and customer service organizations.

Automatic call distributors, ACDs, perform the following functions (an ACD setup is shown in Figure 2.6):

- Route incoming calls to the agent that has been idle the longest.
- Route incoming calls to the appropriate agent group based on the telephone number dialed by the customer or by the customer's telephone number.
- If all agents are busy, either hold the call in a queue, route the call to an alternative group of agents or allow callers, after waiting a specified amount of time, to leave a voice mail message for a later call back.
- Provide reports that measure:
  - Productivity of individual agents on both outgoing and incoming calls.
  - Usage on individual trunks so that managers will know if they have the correct number of lines and that all of the lines are functioning.
  - The number of callers that abandon their calls instead of waiting for an agent.

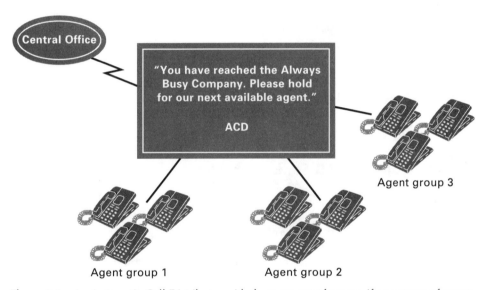

Figure 2.6 An Automatic Call Distributor with three groups of agents. If one group of agents is busy, the ACD can send calls to the second and third groups.

Callers to ACDs can recognize when they reach an ACD if they hear a message such as, "All of our agents are busy, please hold and the next available agent will take your call." Departments that use ACDs typically have more calls than people to handle them. The ACD holds the calls in a queue until someone is available to take the call. Having a machine transfer a call rather than an operator results in savings of about 20 seconds per call. Another saving with automatic call distributors is the consolidation of multiple small groups of agents into fewer large groups. Consider local telephone companies. NYNEX, now Bell Atlantic, previously had small business offices scattered throughout Massachusetts. To save money, Bell Atlantic consolidated its business offices into a few main locations to serve the entire state.

Doubling staff size means more than doubling the number of calls that can be handled in a given hour. This is analogous to the U.S. Post Office and many banks using one long line for tellers and postal clerks rather than an individual line for each "agent." Anyone in a retail store with individual lines knows that getting through these lines takes forever. One always picks the line with the longest wait. With one line for all clerks, it is likely that a clerk will free up more quickly from the "pool" and everyone will get through the line faster.

ACDs are highly profitable for vendors of telephone systems. They are offered both as stand-alone systems and as specialized software and hardware with PBXs and key systems. The main vendor of stand-alone systems is Aspect Communications. These systems are aimed at larger call centers such as those in the airline and financial services industries. Nortel and Lucent have the largest share of PBX-based systems. Most of the larger key systems now offer ACD functionality. Some of them, such as Nortel and Comdial, put their name on third-party systems that they sell with their key systems for small customers who want ACD functionality.

LAN/PBX/ACD Connectivity to Enhance Productivity

Downloading ACD Statistics to PCs

In an ACD environment, everything is measured. Reports on real-time status of calls are the life blood of an ACD. Downloading statistics from an ACD to a PC is commonly done via a local area network. Putting the ACD on a LAN allows information on calls to be manipulated in PC programs such as Excel. Moreover, specialized scheduling programs are designed to run on PCs. These scheduling programs use historical ACD data, often stored on PCs, to determine the correct number of employees to answer telephones each half-hour of the day and during holidays, busy seasons and snowstorms. With connectivity to a LAN, this information can be forwarded electronically to direct supervisors, managers and high-level executives.

LAN to PBX and ACD connectivity is most popular in large call centers. Computer and LAN to ACD/PBX connectivity is used to:

- Download ACD reports to PCs.

- See the real-time status of incoming and outgoing ACD calls and agents on PC screens.

- Improve employee productivity and service to callers.

Voice Response Units—Using the Telephone as a Terminal

Voice response units enable customers to complete transactions without using a live agent, in essence, off loading data entry from agents to callers (see Figure 2.7). Voice response units use DTMF tones (touch tones) to interact directly with computers. For example, when a person calls his or her bank to find out his or her bank balance or credit card company to learn if a payment has been received, he or she no longer has to speak with a live person. Rather, he or she merely enters his or her account number when prompted by an electronic voice and perhaps a personal identification number. The voice response unit then "speaks" back the requested information. These applications enable organizations to provide around-the-clock information to callers without having to pay overtime wages. The following are examples of applications where voice response technology is used to save employee time:

- Cable television, to select pay-per-view movies.
- Newspapers, to enable people to stop and start papers for vacations and report non-delivery of newspapers.
- Mutual fund companies, for trades and account balances.
- Banking, for account transfers and account balances.
- Movie theaters, to play the time and location of movies.
- Within organizations, so that employees can hear about health and pension benefits.
- Universities, for registration and grade reporting.

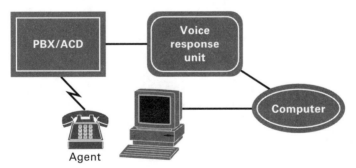

Figure 2.7 A voice response unit connected to an ACD or PBX.

Computer Telephony Integration (CTI)—Routing Callers More Intelligently

Knowing the name of the person or organization calling helps organizations handle calls more quickly. For example, an office supply company can route first-time callers to special agents so that the caller will not have to wait and agents specially trained to interact with new customers can handle the calls.

The telephone number of the caller can be delivered to an ACD in two different ways:

• Directly from the telephone carrier, along with the call.
• Via voice response systems, where callers have been prompted and have entered digits from their telephones.

Special network services, such as ISDN or caller ID, are required to deliver the calling number via the public network. The network service sends either ANI, automatic number identification on interstate calls, or calling line ID. ANI, automatic number identification services work in conjunction with toll-free calls. The ANI is sent by the local telephone call on every toll-free call. In order to receive the calling party's telephone number along with the call, telephone systems need ISDN services. (ISDN is covered in Chapter 7.) Calling line ID is sent on specially equipped local lines. However, this service is generally not used by call centers. Voice response units are an alternative method to deliver a calling party's telephone number.

Voice response units, for example, may ask a person calling to enter, from his or her telephone, his or her account number or telephone number. The number input by the caller is sent from the voice response unit to the organization's computer requesting that information on that account number be sent to the terminal of the agent answering the call. If ANI is used, the telephone system sends the calling party name to the computer system. Having the account information on the screen when the telephone call arrives saves from 10 to 20 seconds per call.

Whether using a voice response unit or automatic number identification, ANI, a third-party integrator must develop software that can understand both the telephone system and the LAN commands and translate between the two (see Figure 2.8). For example,

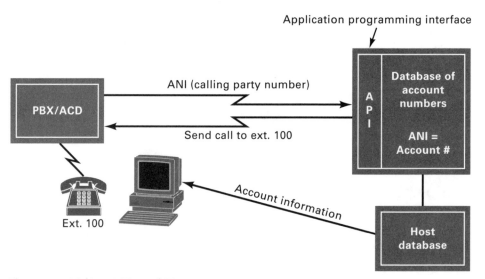

Figure 2.8 Linking ACDs and PBXs to computers.

the computer system may tell the telephone system, route this call to agent A. The telephone system may tell the computer system, agent A is busy on a call. The software used to translate computer to telephone and vice versa is called an application programming interface, or API. The cost for the system integration between the ACD/telephone system and computer system is such that only the largest organizations have implemented these CTI applications. As a matter of fact, many organizations have purchased computer telephony packages and have not implemented them or have only used them for short periods of time due to the difficulty and expense of implementation.

Media: Fiber and Unshielded Twisted Pair Copper

Media is the "stuff" on which voice and data transmissions are carried. Characteristics of media have a direct bearing on the speed, accuracy and distance at which traffic can be carried. For example, thin copper lines carry data more slowly than thicker, higher-quality copper. Fiber carries calls further, faster and with fewer errors than high-quality twisted pair copper cabling.

The two most prevalent media within organizations are fiber optic cabling and unshielded twisted pair (copper). The faster and more expensive of the two choices is fiber optics. The higher cost of fiber optics provides the benefits of higher speed and larger capacity. According to the Corning Web site, "Two strands of fiber can carry more information than a bundle of copper wires four inches in diameter! Two optical fibers can transmit the equivalent of 24,000 telephone calls at one time!" The reason for fiber's superior performance is the fact that it is a non-electric medium.

Another medium being discussed for in-building local area networks is wireless. Wireless has advantages in the inherent mobility it provides. For example, wireless LAN connections enable people to move their computers and reconfigure office space without requiring LAN administrators to physically move cabling. Wireless also has the advantage of carrying data at higher speeds, 100 megabits and greater, which is faster than some existing copper cables. However, wireless is more expensive than unshielded twisted pair to each device and has the added problem of interference in some office environments. One organization, whose facility has one-foot-thick brick walls is not able to use wireless connections because the thick walls inhibit data from traveling through them.

Electrical Properties of Copper Cabling

Electrical properties of copper cabling create resistance and interference problems with transmissions. Sending voice or data over copper cabling is analogous to sending water through a pipe. The further water travels, the weaker it becomes. Similarly, signals weaken the further they are transmitted via copper. Because it is electric, resistance within the copper media slows down the signal or flow of current. The electrical property of copper cabling is the key factor that limits the speed of transmission on copper lines.

Another factor that limits the use of twisted copper wiring is electrical interference. Signals sent over copper wire are direct current electrical signals. Any signals near these signals can introduce interference and noise into the transmission. In particular, cabling in manufacturing areas and near copiers, magnetic sources and radio station transmissions all introduce noise to electrical signals. It is not uncommon to hear offices and residential users complain that they can hear a nearby radio station's programming on their telephone calls.

Within homes and business, crosstalk is another example of "leaking" electrical transmissions. In homes with two lines, one line can often hear the faint conversation on the other line if the wires are too close to each other. In this situation, current from one pair of wires has "leaked" into the other wire. One way copper cabling is protected from crosstalk and noise introduced from nearby wires is by twisting each copper wire of a two-wire pair. Noise induced into each wire of the twisted pair is canceled at the twist in the wire. Twisted pair copper cabling is used:

- From telephone sets to PBX common equipment.
- From telephone sets to key systems common equipment.
- From PCs to the wiring closet of a local area network.
- From homes to the nearest telephone company wiring center.

Category Five Unshielded Twisted Pair—Cat 5

Twisted pair cabling and connection components used inside buildings are rated by the EIA/TIA, Electronics Industry Association/Telecommunications. In 1992, these standards were published such that unshielded twisted pair could be used to transmit data within buildings at 100 Mbps. The most common twisted pair cabling within businesses is rated as either category three or category five. Category three unshielded twisted pair is rated as suitable for voice transmission and Category 5 for data. Category 5 is commonly referred to as Cat 5. Organizations typically save money on data communications by laying copper cabling to individual telephones and PCs rather than installing fiber optics to each desktop. They typically install Cat 5 cabling for both voice and data rather than pay technicians to lay one set of cables for data and another set for voice.

Fiber Optic Cabling—High Capacity and High Costs

Fiber optic lines are immune to electrical interference. Signals are transmitted in the form of off and on light signals rather than as electrical signals. No electricity is present in transmissions over fiber optic. Thus, the signals carried on strands of fiber do not interfere with each other. Therefore, fiber can be run in areas without regard to interference from electrical equipment. Other benefits of fiber are:

- *Security:* Resistance to taps; does not emit electromagnetic signals, therefore, to tap fiber strands, the strands have to be physically broken and listening devices spliced into the break. These splices are easily detected.

- *Small size:* Less duct space required; individual strands of fiber are the size of a length of hair. Duct size is particularly significant under city streets where underground conduit is at capacity, filled with old copper cabling.
- *Single conductor fiber:* Weighs nine times less than coaxial cable.
- *Low attenuation:* Less fading or weakening of signals over distances.
- *No sparking hazards* in flammable areas.
- *High bandwidth:* Suitable for new high-speed transmission speeds such as SONET and ATM.

Disadvantages of fiber optic cabling include:

- Termination, components and connector costs are higher than for copper wiring. Specialized equipment is required to terminate fiber cables within buildings. Specialized equipment is also needed to test and splice fiber and to convert electrical signals to light pulses and vice versa.
- More care in handling fiber is needed. In particular, fiber is not as flexible as twisted pair in bending around corners.
- When fiber is brought into buildings from telephone companies or to the curb in residential areas, local electrical power is required. This adds extra expense.
- Specialized technicians, who may be paid at higher levels, are often required to work with and test fiber cabling.

According to Corning Incorporated's *Just the Facts*, published in 7/95:

The first practical fiber systems were deployed by the telephone industry back in 1977 and consisted of *multimode fiber. Single-mode fiber*, a more recent development was first installed by MCI in a long-haul system that went into service in 1983.

Many long distance vendors' networks now are fully fiber. The added expense of fiber is justified in telephone company networks because of their greater capacity and higher speeds. Telephone companies use fiber because they can lay fewer strands of fiber in less space than heavier copper requiring many more pairs to achieve the same capacity as fiber optic cabling. Moreover, signals can travel further, in the range of 30 miles, on fiber without the use of repeaters to strengthen a faded signal. Thus, fewer repeaters are required in fiber installations. The use of fewer repeaters translates into lower maintenance costs; there are fewer repeaters to break down.

Individual organizations often use fiber cabling:

- For the riser, between floors within a building portion of their networks (see Figure 2.9).
- In campus environments, for cable runs between buildings.

These types of runs require the extra capacity provided by fiber. Instead of needing wide conduit (pipes) to carry fat lengths of copper or coaxial cabling between buildings or

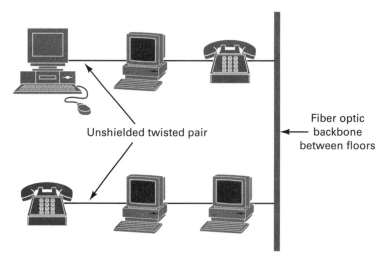

Figure 2.9 Unshielded twisted pair and fiber optic cabling within a building.

floors, organizations lay thin strands of fiber optic cabling capable of transporting vast amounts of voice and data between buildings or floors within a building. The extra expense of laying fiber, however, is usually not justified to individual users and devices on LANs.

Other places fiber is used are:

- In all long distance company networks.
- Between central offices in local telephone company networks.
- From local and long distance phone companies to office buildings.
- From central offices to neighborhood wire centers.
- In Internet service provider networks.
- For undersea cable runs between continents.
- In backbone CATV networks (between the cable company headbands and neighborhood wire centers).
- In electric utility networks.

Fiber is made up of ultra-pure strands of glass over which on and off light pulses are carried. The central *core* of the fiber or its central region is the portion of the fiber strand over which light pulses are carried. The narrower the core, the faster and further a light signal can travel without errors and repeaters. The core is surrounded by *cladding*. The cladding keeps the light contained within the core to prevent the light signal from dispersing, e.g., spreading and losing strength. Finally, there is a *coating* which protects the fiber from environmental hazards such as rain, dust, scratches and snow.

There are two different types of fiber, single-mode and multimode. In single-mode fiber, the combined widths of the core and cladding are as wide as a strand of hair. Single-mode fiber is smaller, more expensive and carries signals faster than multimode fiber. It is used for cable runs longer than two kilometers, or 1.24 miles. This translates to use by telephone company, cable companies, Internet providers and transoceanic cabling. Multimode fiber is used mainly for LAN backbones between buildings on campuses and between floors of buildings.

Single-mode Fiber—Smaller Is Faster and More Expensive

The fact that single-mode fiber is faster than multimode fiber can be explained by the geometric rule: A straight line is the shortest distance between two points.

Picture someone running a race. If this person sticks to a straight line over the course of the race, he or she will get to the end of the race faster than if he or she zig-zagged along the track. This is analogous to single-mode fiber. The small core of the fiber keeps the light signal from bouncing across the diameter of the core of the fiber. Thus, the light signal travels faster than if it had more of a "bouncy" ride through the core. Because it travels more of a straight line, the signal also goes further without attenuation, or weakening. Thus, fewer repeaters are needed. Single-mode fiber can be run for 50 miles without the use of a repeater. In contrast, copper cabling needs to be repeated after approximately one and one-half miles.

While the small size of the core of single-mode fiber enables it to transport data faster, it also increases its expense. The main factor in the increased expense of single-mode fiber is the cost of splicing and connecting it to patch panels and other devices. The core is so small that connections and splices need to be done in a more exacting manner than with multimode fiber. If fiber connections on single-mode fiber do not match cores exactly, the light will not be transmitted from one fiber to another. It will leak or disperse out of the core before being transmitted to the spliced cable.

Components of Fiber Optic Systems

When fiber optic networks are interfaced with devices that transmit electrical signals, the electrical signals must be converted to light pulses. For example, connecting fiber to copper lines or electrical computers requires converters called transmitters and receivers. Transmitters, or light source transducers, convert electric signals to light pulses and receivers, or light detector transducers change light pulses back to electrical signals. Transmitters in fiber optic systems are either LEDs, light-emitting diodes, or lasers. Lasers send faster pulses and are more expensive than LEDs. LEDs are commonly used with multimode fiber.

At the receiving end, the detectors that change light pulses into electrical signals are either PINs, positive intrinsic negative, or APDs, lavalanche photodiodes. LEDs and PINs are used in applications with lower bandwidth and distance requirements. To increase capacity, multiplexers are used in conjunction with fiber transmissions. The multiplexers enable Gigabit speeds and multiple applications to share fiber strands. Once a length of fiber is in place, upgrades in multiplexing technology allow the embedded fiber to carry information at higher speeds.

Industry Overview

CHAPTER 3

Local and Long Distance Providers

The telecommunications industry is in a state of flux created by the passage of the Telecommunications Act of 1996 and the globalization of the worldwide economy. The Telecommunications Act of 1996 allows all telephone carriers to sell both local and long distance calling. To compete in new markets and protect their own territories from encroachment by competitors, local and long distance companies are creating strategic alliances and mergers with companies that have complementary skills and/or additional capital to finance expansion. This chapter presents a detailed chart of the major mergers and acquisitions of the 1990s, as well as an overview of telecommunications suppliers.

Globalization of the worldwide economy is a key factor in the changes occurring in the telecommunications industry. Large American conglomerates who expand their operations to Europe, Africa, Asia and South America do not want the hassle of dealing with a different vendor in each country in which they do business. They want to purchase services from a small number of carriers who can handle their needs for telecommunications services abroad. Moreover, these companies are finding that the rules of providing telecommunications connections differ between countries. A telecommunications vendor that offers the expertise to place orders for telecommunications services in multiple foreign sites has large appeal to multi-national customers.

In addition to sales to global corporations, opportunities for telecommunications sales exist in developing countries in South America, and parts of Asia and Africa. There is a dearth of telecommunications products available in many of these countries. According to Walt Tetschner, President of Tern Systems in Acton, Massachusetts, Venezuela, Colombia and Argentina each have lists of over 500,000 people waiting for telephone lines. India alone has a waiting list of over two million people. This contrasts to the U.S. where the rate of growth of long distance services is slowing down. In addition, in many countries of the world, telecommunications competition is being allowed for the first time. In many places, governments previously allowed monopolies such as PTTs, Post Telephone and

Telegraphs, to control telecommunications services. This is changing, markets are opening up and carriers are eager to gain market share in these newly opened markets.

To give them the expertise and capital to expand their global reach, organizations are creating alliances and mergers with foreign partners. For example, in July of 1997, AT&T announced a partnership with Stet, the national carrier of Italy. Together, they plan to develop markets in South America for data and voice telecommunications services. Both Stet and AT&T have existing customers in South America. An example of a merger is the British Telecom purchase of 20 percent of MCI. This alliance created an international venture called Concert Global Communications PLC to sell services internationally. This is a case, where one partner, British Telecom, the telephone company in England, brought capital and presence for worldwide expansion. Concert sells services in 72 countries.

In anticipation of increased competition in local and long distance calling legislated by the Telecommunications Act of 1996, vendors are creating strategies to expand the services they sell. Between 1984 and the mid-1990s, Bell telephone companies such as Bell-South, US West and Ameritech sold mainly local and local toll services, yellow page advertising and cellular services. Interstate carriers such as AT&T and MCI sold mainly interstate voice and data telecommunications services. These organizations now provide both local and interstate calling services. The Regional Bell Operating Companies are also expanding overseas. For example, NYNEX sells cable TV service in England. As a result of the Telecommunications Act of 1996, differences between local and long haul providers are becoming less marked.

To compete in a broader range of markets, local telephone companies are merging with corporations that have complementary strengths and additional capital for expansion into new markets opened up by the Telecommunications Act of 1996. The new markets that Regional Bell Operating Companies and independent local telephone companies such as GTE and Frontier Telecommunications plan to participate in are interstate long distance from within their regions and interstate long distance originating in states outside of their home territories and Internet access. They are also growing by adding new cellular and paging services. An example of a local telephone company merger is the Regional Bell Operating Company SBC's purchase of another Regional Bell Operating Company, Pacific Telesis. Additionally, Bell Atlantic merged with NYNEX. Finally, GTE, the largest independent non-Bell telephone company, purchased BBN, an Internet network and access supplier.

As local Bell and independent telephone companies prepare to compete against them, interstate carriers are also looking at new markets in which to expand. Interexchange carriers such as AT&T, MCI and Sprint have an increasing share of wireless and Internet access services. Both AT&T and Sprint are leading providers of cellular services. Sprint sells cellular services with its cable company partners, Comcast, TCI and Cox Communications, under the Sprint Cellular name. AT&T is the largest cellular provider in the U.S. having purchased McCaw Cellular. MCI has a large Internet network. The point is that local and interexchange carriers are expanding into other markets here and abroad.

In addition to local and interstate carriers, there are groups of vendors who resell long distance and local services. Resellers, aggregators and agents resell interexchange carrier and local telephone company calling services. Resellers purchase bulk services from carriers such as AT&T, MCI and Sprint at discounts which they pass on to customers who often don't have the market clout to obtain discounts directly. In the 1980s and 1990s, indirect as well as direct sales channels of distribution were developed. Indirect channels, e.g., resellers, aggregators and agents, are experiencing considerable growth in the 1990s. They aggressively market their services to customers who are often too small to merit a great deal of time from large carriers and local telephone companies.

Table 3.1 lists the major mergers and alliances announced in the early to mid-1990s.

Table 3.1   Mergers and Acquisitions in the Telecommunications Industry

Purchasing Company	Purchased Entity	Details
AirTouch	New Vector Group, Inc. from US West	Announced in April of 1997. New Vector was the cellular arm of US West. AirTouch had previously purchased the cellular telephone portion of Pacific Telesis. This makes AirTouch the second largest wireless company behind AT&T Wireless.
AT&T	BBN Planet	AT&T purchased a minority stake. Gave AT&T Internet service provider capabilities. At the time of the GTE purchase of BBN, BBN carried AT&T WorldNet's Internet traffic.
	Interchange Service	Interchange service, a planned on-line network, was owned by Ziff-Davis.
	McCaw Cellular Communications	Provided AT&T a foothold into wireless services.
	Teleport Communications group	Announced January of 1998 Teleport sells local telephone service in 66 cities over its own fiber lines. This gives AT&T a foothold to sell local telephone service.
Bell Atlantic	NYNEX	A merger of two Regional Bell Operating Companies.
British Telecommunications PLC	20% of MCI	British Telecommunications PLC purchased 20% of MCI in 1993. Market jointly internationally under the name Concert. Merger may not be completed due to October 1997 higher bid for MCI by WorldCom.

Table 3.1   Mergers and Acquisitions in the Telecommunications Industry *(Continued)*

Purchasing Company	Purchased Entity	Details
Deutsche Telekom, France Telecom	Sprint Corporation	20% stake in Sprint for the purpose of joint efforts in the international telecommunications market. New venture called Global One.
GTE	BBN	GTE, the largest independent telephone company, announced its intention, May 6, 1997, to purchase Internet network provider BBN. Gives GTE the capability to provide a full spectrum of services from local, high-speed data and long distance to Internet services.
MCI	Nationwide Cellular	Nationwide Cellular operated cellular services in ten cities. Became the basis of NetworkMCI cellular services division.
	SHL Systemhouse	SHL was a Canadian network management company. SHL is now called MCI Systemhouse. It specializes in outsourcing.
	Western Union Corporation's Advanced Transmission Systems Division	Gave MCI a jump-start in building fiber optic lines in cities for the purpose of selling local calling services. Now part of MCI Metro.
MFS	UUNetTechnologies	Gave MFS, a competitive access provider, an Internet service provider.
Net's, Inc. (formerly Industry Net. Net's, Inc. declared bankruptcy in May of 1997).	AT&T New Media Services	AT&T sold Interchange service, whose name they had changed to AT&T New Media Services to the founder of Lotus' Industry Net. AT&T has a partial ownership in Net's, Inc., an on-line listing of buyers and sellers of industrial goods.
SBC Communications, Inc. (Southwestern Bell Communications)	Pacific Telesis Group	A merger of two Regional Bell Operating Companies.

Table 3.1 Mergers and Acquisitions in the Telecommunications Industry *(Continued)*

Purchasing Company	Purchased Entity	Details
Science Applications International Corp., SAIC	Bell Communications Research, Bellcore	Announced 9/96 pending approval by all Regional Bell Operating Companies, original owners of Bellcore. SAIC owns the InterNIC.
Sprint	Centel Cellular	Became Sprint Cellular in 1993, spun off in 1995 to become an independent company named 360°. Sprint Cellular now sell PCS cellular with Cox Communications, Comcast and TCI.
US West Media	Continental Cable	Provided a source of cash for Continental Cable to expand its service offerings. New venture called MediaOne.
WorldCom, Inc.	Brooks Fiber Properties	Announced in 1997. If finalized, will add more local fiber optic networks.
	CompuServe Corp.	WorldCom will keep CompuServe's 1200 corporate Internet customers and turn over CompuServe's three million residential customers to AOL. AOL, in turn, will transfer its Internet access and backbone to WorldCom. This purchase, if approved by the U.S. government makes WorldCom the largest provider of Internet services.
	MCI	Announced in October 1997. If finalized, will provide more Internet backbone and local fiber optic networks as well as interexchange facilities to WorldCom.
	MFS	Provided WorldCom, the fourth largest interexchange carrier, with local fiber optic networks and Internet service provider services.
	WilTel, IDB, Metromedia	Enabled WorldCom to become the fourth largest domestic interexchange carrier.

In addition to resellers, competitive access providers such as MFS and TCG enlarged their scope of business activities during the 1980s and 1990s. Competitive access providers, CAPs, initially provided fiber optic links between customers in major metropolitan areas and interstate long distance providers. The purpose of these links was to connect long distance vendors to local customers without the hefty access fees charged by local telephone companies. Having fiber optic links in place in major metropolitan areas has enabled CAPs to expand their offerings. CAPs now sell local and interstate calling services, data communications services and Internet access.

An additional area of growth is access to the Internet. Major players in both local and long distance services are adding Internet access to their portfolios of offerings. To do this, they are buying up existing providers. AT&T purchased a stake in BBN Planet, and MFS, an alternate access provider, bought UUNetTechnologies. Both BBN Planet and UUNetTechnologies are Internet access providers.

The bottom line is that many key players want to offer a full spectrum of services. These offerings include Internet access, local toll calling, interstate toll calling, wireless calling, satellite services and delivery of entertainment services. Consider Robert Allen's, (chairman of AT&T) statement in a press release dated February 8, 1996, the date the Telecommunications Act of 1996 was passed:

> We will offer business and consumers bundles of services that will combine local and long distance, wireless, on-line services, even television. As much or as little as the customer wants.

The Bell System Prior to and after 1984

Before 1984, the Bell system consisted of 22 local Bell telephone companies that were owned by AT&T. AT&T and the Bell companies sold local, domestic U.S. and international long distance services, as well as customer premise telephone hardware. *Customers had one point of contact for all of their telecommunications requirements.* AT&T with its 22 Bell Operating Companies:

- Sold local, interstate and international long distance.
- Manufactured and sold central office switches, customer premises telephone systems, electronics and consumer telephones.
- Provided yellow and white page telephone directories.

This monopoly by AT&T on both interstate and local telephone facilities hampered competition in long distance services. For example, companies such as MCI and Sprint that wished to compete against AT&T in a metropolitan area such as New York City, needed connections provided by the AT&T subsidiary, New York Telephone Company, to complete calls to New York residents.

Also, consider the following: An interexchange long distance company, such as Sprint, owned network facilities between Boston and New York City on which it carried calls. It had no lines to individual homes and businesses in, for example, the New York area. To complete and originate calls from businesses and consumers, interexchange carriers, such as Sprint, needed connections to the local telephone company. The lines and central office facilities owned by a local Bell telephone company such as New York Telephone were required to carry calls from an individual subscriber to Sprint's long distance facilities. Calls could not leave or enter the interexchange carriers' networks without connections from local telephone companies. These local telephone companies at this time, prior to 1984, were owned by the very organization, AT&T, with which the new carriers were competing. Access from AT&T's local telephone companies to competitive carriers such as Sprint is illustrated in Figure 3.1.

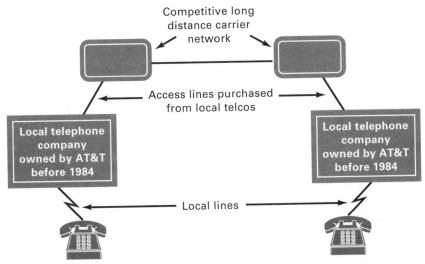

Figure 3.1   Local access interconnection.

By 1974, so many complaints had been filed with the Justice Department about AT&T's lack of cooperation in supplying connections to local phone companies that the Justice Department filed an antitrust suit against AT&T. In 1984, the suit was resolved. The Justice Department divested AT&T of the 22 local phone companies. The resolution of the Justice Department case against AT&T is known as the Modified Final Judgment or divestiture. Ownership of the 22 local phone companies was transferred from AT&T to seven Regional Bell Operating Companies. The seven Regional Bell Operating Companies, RBOCs, at that time were (the indented companies are the former 22 Bell telephone companies):

- Ameritech
    - Michigan Bell Telephone Company
    - Ohio Bell Telephone Company
    - Indiana Bell Telephone Company
    - Illinois Bell Telephone Company
    - Wisconsin Bell Telephone Company

- Bell Atlantic
    New Jersey Bell Telephone Company
    Bell Telephone Company of Pennsylvania
    Diamond State Telephone Company
    Chesapeake and Potomac State Telephone Company of Virginia
    Potomac State Telephone Company of Maryland
    Potomac State Telephone Company of West Virginia
    Potomac State Telephone Company of Washington, D.C.
- BellSouth
    Southern Bell Telephone Company
    South Central Bell Telephone Company
- NYNEX
    New England Telephone Company
    New York Telephone Company
- Pacific Telesis
    Pacific Bell Telephone Company
    Nevada Bell Telephone Company
- Southwestern Bell Communications (SBC Communications)
    Southwestern Bell Telephone Company
- US West
    Mountain Bell Telephone Company
    Northwestern Bell Telephone Company
    Pacific Northwest Bell Telephone Company

The seven RBOCs, Regional Bell Operating Companies, retained the "Bell" logo and the right to sell local and toll calling within the local area. They also retained the lucrative white and yellow page markets. However, they were denied the right to manufacture equipment. A centralized organization, Bellcore, owned jointly by the RBOCs, was formed. This centralized organization had two functions. It was a central point for National Security and Emergency Preparedness and also a technical resource for the local telephone companies. As part of its technical resource services, it administered the North American Numbering Plan. For example, it coordinated allocation of area codes.

AT&T retained the right to manufacture and sell telephone and central office systems and to sell interstate and international long distance. AT&T kept Western Electric for manufacturing and Bell Labs, its research arm.

Bell and Independent Telephone Companies

From 1984 to 1996, the Bell companies sold basic services such as local and local long distance services. On March 7, 1988, U.S. District Court Judge Harold Greene lifted some line of business restrictions against the Bells. He allowed them to offer enhanced services. Enhanced services employ computer processing to act on subscriber transmitted information. Examples of enhanced services are voice mail, audiotext (press 1 to hear about skiing conditions, etc.) and electronic mail services. In exchange, the Bell System opened up more than 100 network features to competitors. These features included call forwarding to competitive vendors' voice mail and reconfiguration of certain lines by customers.

In 1996, the Telecommunications Act of 1996 was passed, which opened the Bell territories to further competition from long distance vendors, cable companies, local access providers and utility companies. In anticipation of the requirement for major capital to expand into markets opened by the Telecommunications Act of 1996 and to compete in their own territories more effectively, four of the above RBOCs announced mergers in 1996. The announced mergers were Pacific Telesis with SBC and NYNEX with Bell Atlantic. Figure 3.2 illustrates the areas encompassed by each RBOC. These mergers decreased the number of RBOCs from seven to five:

- Ameritech.
- Bell Atlantic with NYNEX.
- BellSouth.
- Southwestern Bell Communications, (SBC Communications) with Pacific Telesis.
- US West.

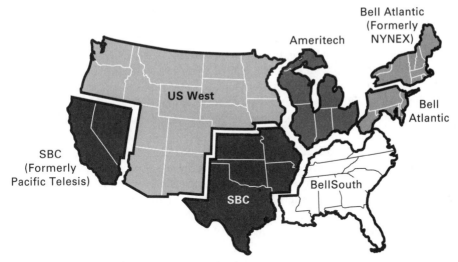

Figure 3.2   Bell Telephone company territories.

The markets opened to the Bells by the Telecommunications Act of 1996 were manufacturing of telephone systems and the sale of interstate long distance both within their regions and throughout the country.

The Bell companies' greatest revenue is generated from selling in-state telephone service, interconnecting interexchange carriers to local service, providing transport of cellular calls and publishing white and yellow page directories. Ameritech states in their 1995 annual report:

Mergers and Partnerships

The need for cash to enter new local and long distance markets has led to mergers, including: Continental Cable with US West Media, LDDS WorldCom with MFS, Bell Atlantic with NYNEX and Pacific Telesis with SB Communications.

In addition, RBOCs are partnering with carriers such as LDDS WorldCom and Frontier Telecommunications. In this way, they can sell long distance outside of their territories without building new networks.

To illustrate, Ameritech is selling long distance services to its cellular customers in its home regions of Illinois, Missouri, northwest Indiana, Ohio, Wisconsin and Michigan. Calls out of the region will be transported on WorldCom's network.

Our core business-telephone, cellular and directories is growing...we strengthen our core business by adding new services such as security monitoring, cable TV and long distance. We reach customers around the United States, Europe, Asia and the Pacific Rim through...alliances, strategic investments and export of services.

Ameritech's 3.7 million cellular customers in Hungary and New Zealand are an example of one Bell telephone company's new international service. The Bells are also expanding aggressively into supplying entertainment services. For example, Time Warner has a partnership with US West; NYNEX has part ownership of Viacom and has also formed an entertainment programming company called TeleTV; and Sprint has partnerships with cable companies Cox Cable Communications, Tele-Communications (TCI) and Comcast. However, as stated above, the Bells' core business is local telephone service

In addition to the RBOCs, there are close to 1400 independent telephone companies. Some of these are GTE, Frontier Telecommunications and Southern New England Telecommunications (SNET). Many of the independent telephone companies are in rural areas such as northern Maine and western Massachusetts. According to the telephone company industry association, USTA, United States Telecommunications Association, independent telephone companies supply dial tone to 15% of the telephones in the U.S. but half of its geographic area.

Independent telephone companies sell all of the same services that RBOCs sell. In addition, following the 1984 divestiture, they were allowed to directly sell equipment such as telephone systems. This contrasts with the Bell telephone companies, who were not allowed to sell telephone equipment, only telephone calling services.

Impact of the Telecommunications Act of 1996

The objective of the Telecommunications Act of 1996 was to open the lucrative, $80 billion per year in revenue, local telephone service to deeper competition. In addition, it opened the interstate long distance market to competition from local telephone companies.

The Telecommunications Act of 1996:

- Permitted RBOCs to sell in-region long distance after completing a 14-point checklist of how they would offer connection into local calling for long distance companies such as AT&T. The 14-point checklist was designed to prove that there were alternative sources for local competitive calling.

- Freed interexchange carriers, CAPs, cable companies, wireless service operators, broadcasters and gas and electric utility companies to sell local telephone services.

- Required local telephone companies to offer resale of and interconnection to local services to the above entities.

- Authorized local telephone companies to sell cable services, television services, equipment and out-of-state long distance, voice messaging and cellular services.

- Raised the limit on the number of TV stations networks could own and phased out cable TV rate regulation.

- Promised carrier reimbursable discounts to schools, health care institutions and libraries in rural areas for access to advanced telecommunications services.

- Allowed RBOCs to manufacture goods through separate subsidiaries.

Interexchange Carriers

Prior to the Telecommunications Act of 1996, interexchange carriers sold long distance services primarily between states and to international locations. Examples of interexchange carriers are:

- AT&T.
- MCI.
- Sprint.
- Cable and Wireless, Inc.
- LDDS WorldCom.
- Frontier Telecommunications.
- Allnet.

Interexchange carriers, IEXs, own most of the switching and transmission equipment over which the traffic they bill for is routed. For example, they own fiber optic cabling, microwave towers, multiplexing equipment (to send multiple voice and data conversations over the same fiber cable) and telephone equipment (switches) that routes calls.

Services they sell include:

- Toll-free 800 and 888 services.
- Outgoing long distance.
- Dedicated private lines.

MCI: How New Technology Lowered Barriers to Entry into the Long Distance Market

MCI, originally known as Microwave Communications, Inc., got its start with the invention and improvements in microwave. Microwave technology was heavily invested in during World War II and improved at the Massachusetts Institute of Technology. Further development took place in the 1960s. MCI successfully used microwave technology to implement private networks, originally between Chicago and St. Louis. The use of microwave technology eliminated the expense and labor intensity of laying copper cables. It was a key factor in opening competition to AT&T by lowering the cost of building telecommunications networks.

- Local calling services.
- Data transmission services.
- PCS cellular services.
- 900 services.
- Internet access.
- Cellular wireless services.

In addition to their interest and presence in local and long distance calling, interexchange carriers are exploring the transmission of entertainment services. AT&T, tried through its previous financial stake in DirecTV, promoting the sale of satellite dishes to its long distance customers. The satellite dishes carried sports, movie and music programming to peoples' homes. Sprint has partnerships with cable companies Comcast, TCI and Cox to deliver entertainment and telephone services to local customers.

Moreover, the Telecommunications Act of 1996 opened the local telephone companies as both competitors and customers. While the Bells will be able to sell interstate long distance services, they do not have networks in place to carry these calls. Interexchange carriers' lines are the vehicles for transporting the interstate calls sold by Bell telephone companies. The interstate carriers are in heavy bidding wars for Bell resale contracts. For instance, LDDS WorldCom has an agreement to carry Ameritech's cellular long distance calls.

Transporting Calls between Local Phone Companies and Interexchange Carriers

The conditions imposed by the 1984 divestiture agreement between AT&T and the U.S. Justice Department limited interexchange carriers to providing long distance services between LATAs. A LATA, or local access and transport area, is defined as an area within which an RBOC can deliver calls. LATA boundaries, for the most part, fall within states. Interexchange carriers were barred from carrying calls within LATAs. All calls *within* LATAs were to be carried by local phone companies. All calls *between* LATAs were to be carried by interexchange carriers. (See Chapter 5 for details on LATAs.)

Thus, the interstate, or more correctly, the inter-LATA portion of calls was transported by interexchange carriers and handed off to local phone companies to be carried to their final destinations. This hand-off of calls to local exchange carriers is the local access portion of inter-LATA calls.

Prior to the 1996 Telecommunications Act, transporting calls from the interexchange portion of the public network to homes and businesses was analogous to taking a trip by airplane. Airplanes transported riders from airport to airport. Taxis, limousines and buses transported passengers from airports to their homes and businesses. Airplanes are the interexchange, or inter-LATA portion of the call, and taxis and buses are the local telephone company portion of the call. Interexchange carriers, like airplanes, picked up and handed off calls at airports to local "transportation," the local telephone company (buses and taxis in this analogy).

This local transportation that carries calls from interexchange carriers' "airports" was defined in the 1984 divestiture as belonging to local exchange carriers. In telecommunications parlance, the airport, or interexchange carrier "drop-off" and "pick-up" points, became known as the POP. POP is short for point of presence. It is the location of the interexchange company's telephone switch that connects to the local telephone company. Interexchange carriers rented lines and central office connections from their POPs to the local telephone central offices to transport calls to and from their POPs (see Figure 3.3).

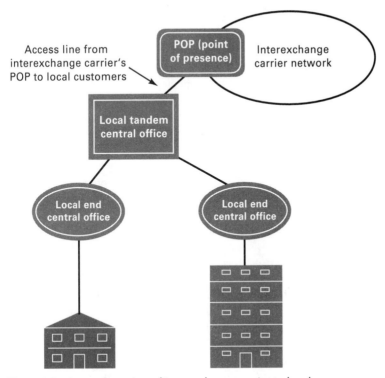

Figure 3.3 Interconnection of interexchange carrier to local exchange carrier.

Each major interexchange carrier has a point of presence, or POP, in each metropolitan area. For example, AT&T has two POPs in the Boston area, one in Cambridge and one in Framingham. Every business and residential customer that used AT&T for long distance, prior to the Telecommunications Act of 1996, had to have "transportation" from his or her location to the carrier's POP. The requirement for local access and egress is a big factor in the development of another type of telecommunications services provider, the competitive access provider (CAP).

Competitive Access Providers (CAPs)

The cost to carry the local access portion, the "taxi or bus" access and egress (exit), was paid by interexchange carriers to local telephone companies and passed on to the consumer. Access charges amounted to approximately 4 1/2¢ per minute. Bell companies derived about one-third of their revenue from interexchange carriers' access fees. AT&T was the largest customer of most Bell telephone companies. Many Bell Operating Companies received a third of their revenue from business customers, a third from residential customers and a third from access fees paid by interexchange carriers.

In the early 1980s, business customers became aware that these access charges were costing them large amounts of money. Moreover, it was taking a long time for connections to long distance companies to be installed. In 1983, Merrill Lynch formed Teleport Communications Company to carry its calls from Merrill Lynch's New York City locations to its interexchange carriers. It thus bypassed New York Telephone Company access fees. This was the beginning of the competitive access provider (CAP) industry.

The cost to transport calls to and from interexchange carriers' networks to their end destinations in local cities was a key factor in the initial development of competitive access. CAPs, such as Metropolitan Fiber System, MFS, Intermedia Communications of Florida, Inc. and Teleport Communications Group, TCG, are also known as alternative access providers and local access providers. They supply "alternatives" for local phone service and access to interexchange carriers from local customers.

Advances in fiber optics and digital transmission of telephone calls played a key role in the introduction of competitive access. The CAPs used fiber optic cabling to transport calls from business customers (initially, primarily financial institutions in large cities) to interexchange carriers. The local telephone companies at this time, the early to mid-1980s, were transporting calls from customer premises on copper cabling. (Fiber is more reliable and carries calls with fewer errors. It is a more reliable medium.)

The digital connections used in the 1980s were T-1. T-1 carries 24 channels of voice and/or data calls on two pairs of copper cable or one pair of fiber cabling. These digital links between interexchange carriers' points of presence, POPs, and business customer premises enabled customers to "bypass" the 4 1/2¢ per minute access fees charged by the Bell telephone companies. Bypass was the main springboard from which the CAPs entered into additional telecommunications services.

This competition from alternative access providers pushed the Bell telephone companies to lay more extensive fiber in large cities, directly to customers' buildings. It also

Central Office Fire Leads to Founding of MFS, Metropolitan Fiber Systems

MFS was founded in 1988 by Peter Kiewit Sons, a construction company based in Omaha, Nebraska. Their main business was building fiber optic networks for carriers such as AT&T and Sprint.

The central office fire of 1987 triggered MFS' concept of bypassing the local exchange company. The fire knocked out all telephone service in the Hinsdale exchange for two weeks starting on Mother's Day weekend. One organization which lost service was FTD, whose main center was in Hinsdale. As a result of the fire, FTD was not able to take calls for flower deliveries.

The fire raised the notion of back-up to the local telephone central office: "What happens if the central office goes down?" The fire created an opportunity for an industry. By using a competitive access provider, CAP, for connection to their long distance vendor, customers could retain long distance services in the event that the central office crashed. MFS built their first bypass network in Chicago in unused underground coal tunnels.

pushed the Bells to shorten intervals for high-speed digital access to long distance carriers. Customers had choices of how they transported their calls from their premises to the carriers' POPs. They could use plain old telephone lines from the Bells' central offices to the POPs, T-1 from CAPs or T-1 from the Bells themselves. Faster implementation, lower costs and reliability led many institutions to select a CAP. In response, the telephone companies more quickly than they had planned, made digital, T-1 connections over fiber available. Thus, competition from alternative access providers was a factor in the speed of upgrading the Bells' cabling from copper to fiber.

CAPs sell the following:

- Fiber optic cables connecting commercial and business customers to interexchange carriers' switches, POPs (see Figure 3.4). This direct connection from end-users to long distance carriers "bypasses" the local telephone company's access charges.

- Connections to the Internet for both end-users and Internet service providers. Internet service providers resell Internet access to businesses and individuals.

- Sophisticated data networks within metropolitan areas: Firms that have, for example, both a warehouse and headquarters in the same metropolitan area may want a high-capacity connection between the two locations to send inventory and/or sales orders between the two sites.

- Local telephone numbers (dial tone) and local calling services: Once the CAPs have their own fiber in a metropolitan area, the next step is to add switches and begin offering local telephone service.

Local access suppliers have expanded their lines of business from local access and private lines, to local dial tone, interexchange long distance and Internet access. They are

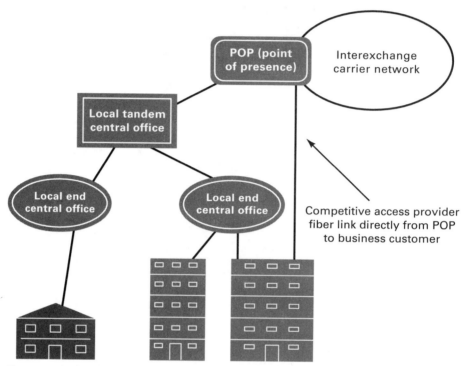

Figure 3.4   Alternative access link to a long distance provider.

becoming full-fledged carriers. CAPs are entering the interexchange business by constructing fiber optic links between major metropolitan areas in the east and west coast corridors. For example, once Los Angeles and San Francisco have fiber within the cities, the next step is to link these two locations for the purpose of selling long distance. The same is true for New York City to Washington, DC. These fiber optic lines will be available for the transmission of voice and data between cities on these high-traffic corridors.

In terms of selling local calling services, CAPs are leveraging their miles of fiber optic cabling into providing full local services. They are installing their own local switches, often purchased from the equipment arm of AT&T, now Lucent Technologies. These switches are used to connect the fiber optic cabling into a vehicle to supply local telephone service. They are in a position to not only connect customers to carriers' POPs, but to actually route calls between business and residential customers in metropolitan areas.

Resellers

Telecommunications services are sold both directly to end-users and through resellers. According to ATLANTIC*ACM, a Boston-based consulting firm, in 1995, *19.2% of long distance revenues were generated by resellers.* Companies that sell to resellers are

---

The Merger of MFS with UUNet Technologies and then WorldCom—Bypassing the Local Phone Company for Internet Access and Having It All

In April of 1996, alternative access provider MFS purchased Internet services provider UUNet Technologies, Inc. MFS has fiber lines in 50 domestic cities and several international cities in Europe and Asia. End-users in these cities will be able to connect directly to the Internet via MFS' fiber to UUNet Technologies' switches.

This arrangement bypasses local telephone companies and a significant portion of Internet access costs. This is another example of competing with the local phone companies. It is also a part of the trend toward the development of one-stop shopping services by enabling MFS to sell local telephone service and direct interconnection to the Internet.

Following this merger, MFS was purchased by LDDS WorldCom in August of 1996. WorldCom itself was created from the mergers with the following carriers: IDB, Metromedia and Wiltel. At the time of the WorldCom/MFS merger, the two companies had combined revenue of $1.3 billion and net income of $500 million. MCI alone had an income of $550 million in 1995. The WorldCom MFS combination creates formidable competition in the telecommunications marketplace for a vendor strong in both local and interstate services.

---

known as wholesalers. AT&T, MCI, LDDS WorldCom, the Regional Bell Operating Companies and Sprint sell both directly to end-users and through various types of resellers.

According to ATLANTIC*ACM, revenue generated by resellers grew by 31% from 1993 to 1995. There are four types of resellers:

- *Resellers:* own switches that route calls; lease lines from major carriers over which the calls are routed.
- *Switchless resellers:* do not own switches; purchase large quantities of switched services which they resell; receive billing tapes from carriers with detail from which they bill customers directly for services.
- *Aggregators:* sign up for large blocks of calling capacity from carriers; pass on these discounts to customers, but do not bill customers directly.
- *Agents:* work as independent sales agents, mainly for Bell telephone companies.

Resellers, switchless resellers and aggregators buy long distance services wholesale and resell them at retail prices to end-users.

Resellers and Switchless Resellers

As the name implies, resellers "resell" long distance services. The services they resell are from major carriers such as Sprint, MCI, LDDS WorldCom and AT&T. Resellers lease network facilities such as fiber cable runs from carriers. Resellers own switches in high-volume calling areas that route calls over the fiber cable lines leased from companies such as AT&T, Sprint and MCI. Resellers do not necessarily have switches in all of the locations from which they sell long distance. They generally place them in high-volume calling areas where they can justify the expense of a switch and eliminate paying the underlying carrier for a portion of the switching and bill tape expenses.

Switchless resellers, unlike resellers, do not own any of their own switches. They purchase and resell services from multiple long distance carriers. The underlying carriers for whom the switchless resellers sell services send billing magnetic tapes to the switchless reseller. The switchless reseller uses these tapes to generate bills which they send to their customers.

Switchless resellers and resellers buy long distance capacity in bulk and resell it to small and medium-sized organizations. They bill customers directly and have their own customer service and sales staff to which customers report service problems. The reseller, in turn, coordinates resolution of problems with carriers. Customers report repair or billing problems directly to the reseller.

Resellers see their role as reaching customers too small to receive large discounts directly from carriers. Carriers such as AT&T view resellers as both competition and as another sales channel to reach smaller organizations. Individual AT&T sales reps compete with AT&T resellers. AT&T corporate, however, use resellers as another distribution channel.

A sampling of resellers:

• Network Plus	Quincy, MA
• US Networks	New York City, NY
• International Telcom Ltd.	Seattle, WA
• Unidial Communications, Inc.	Louisville, KY
• Excel Communications	Dallas, TX
• Global Spectrum Telecommunications Corp.	Fort Lauderdale, FL

With the opening of the local telephone market to long distance vendors, resellers are requesting regulatory approval to resell Baby Bell offerings. For example, US Networks has plans to resell Ameritech and NYNEX calling services and enhanced services such as telephone company voice mail and conference calling. Local toll calling and services are additional sales opportunities for resellers.

The Bell telephone companies thus sell services directly to end-users and indirectly through resellers. In essence, they are both retailers and wholesalers of services. Resellers have the advantage of tapping into their existing long haul customer base when they sell local services. To illustrate, a reseller of interstate long distance can approach its customers in Chicago for resale of toll calling within Illinois and the Chicago metropolitan area. Another form of reseller is the alternate operator service provider, AOS company. AOS companies supply long distance to privately owned pay phones and telephones in public locations such as airports, hospitals and universities. They buy long distance in bulk from interexchange carriers and resell it at a premium. Calls from these phones are intercepted by alternate operator services equipment. Their highly automated equipment validates credit card numbers and keeps track of calls for billing purposes. They also process debit card calls. Examples of AOS providers are Oncor Communications, Amnex, Peoples Telephone Company and Capital Network System.

Pay Now, Hope to Call Later—Debit Cards

Debit cards represent an area of opportunity for resellers, operator service companies, carriers and agents. Debit cards, which were introduced in the U.S. in the late 1980s, were, according to a *Wall Street Journal* article of July 16, 1996, a $1 billion industry in 1995. Bellcore estimates that it will be a $5 billion dollar industry by the end of the century.

Debit cards were first used in Europe in 1976 where credit is not a ready option for calling cards. With debit cards, people without access to credit or without their own telephone service can purchase telephone calling. Users pay for the card in advance and are allowed a set amount of time or calling units. The cards gained popularity in the U.S. initially as a way for people without telephones or credit to make telephone calls. Debit cards are sold to:

- Travelers.
- Foreign and American students who are often not eligible for calling cards.
- Immigrants.
- Hospital patients so that hospitals don't have to bill patients for long distance calls.
- Collectors.
- Businesses who use them for promotional gifts to their customers.

Debit cards are collectible, marketing tools and promotional items. For example, SmarTel, a marketer of debit cards headquartered in Boston, assists organizations in developing promotional programs in conjunction with the distribution of debit cards. In one such program, a cable company awarded cards with 30 minutes of calling time to customers that upgraded to premium channels. The cards could be used either for long distance calling or to hear brief cast and plot summaries of upcoming movies. According to Howard Segermark, Executive Director of the International Telecard Association, debit cards are a popular promotional item because their perceived value is higher than their actual cost.

Debit cards are sold directly by carriers, resellers, sales agents, distributors and retail outlets. The debit card industry is a "layered" industry with different segments selling the cards, promoting the cards, handling the administration and billing aspects and actually carrying the calls. For example, many of the local Bell telephone companies merchandise the cards themselves, but outsource the processing of calls. The cards are often sold by the Bell telephone companies or their agents, but the calls themselves are frequently handled by AOS providers. These alternate operator service providers buy bulk long distance from carriers at wholesale prices. They connect their own switches to the leased, bulk carrier lines.

The "prompts" callers hear when they use debit cards, e.g., "Please enter the PIN number on the back of your card," are generated on the alternate operators' equipment. The amount of elapsed time and allowed time per debit card are also tracked by the operator services companies. The calls themselves are carried by interexchange carriers.

Buyer Beware
Some agents call themselves consultants. They offer, at no fee, to evaluate an end-user's requirements and recommend a cost-effective solution. Buyer beware—there is no "free lunch." Agents make their money by selling equipment and services. The consulting is often a way for agents to uncover applications for new network services and telephone equipment.

Aggregators

Unlike resellers, aggregators do not lease lines directly or own their own equipment and switches for routing calls. Aggregators sign up with carriers such as AT&T for discounts similar to those available to Fortune 100 companies. They pass on to their customers deep discounts gained by promising large volumes of calls to carriers. The aggregator offers a portion of this discount to end-users who are too small for such discounts. Some aggregators bill customers directly, keeping a percentage of the savings for themselves.

Other aggregators have the long distance carrier bill the service directly. In these cases, the end-user reports repair problems directly to the carrier.

Agents

Agents are companies that work on commission for local telephone and interexchange companies. They enter into agreements to sell long distance, local toll calling, cellular service and value-added services such as voice mail, paging and local Centrex services. (See Chapter 2; with Centrex, most of the hardware for business telephone systems is located in the telephone company central office rather than at the customer premise.) Many agents sell telephone equipment, inside cabling and other hardware such as video conferencing systems, as well as resell long distance, cellular and local calling services.

Typically, customers sign letters of agency with the agent when they agree to purchase telephone calling services. The agent places orders for service with the carrier and/ or phone company. The agent also manages the installation and reports repair problems for customers. The customers are then billed directly by the phone company or carrier for services sold by agents. The agent receives a monthly commission from the carrier. In the late 1980s when agents were first in the business of selling local telephone calling, they sold only local toll and value-added services. They subsequently added interexchange calling to their "portfolio" of services.

Frequently, agent programs are successful because of the strong sales efforts made by agents. In contrast, the Bell telephone companies often focus their major sales efforts on large customers. Small and medium-sized customers are courted by agents who sell them telephone systems, as well as voice, cellular and data calling services.

CHAPTER 4

The Telecommunications Act of 1996 and Local Competition

The Telecommunications Act of 1996 addresses in detail the issue of competition in local telecommunications services. It mandates that Regional Bell Operating Companies, who are eager to sell out-of-state long distance services, open up their facilities for connection to competitors. To this end, they are required to make their telephone lines, local switches, repair services, operator services, telephone book listings and databases available to competitors. This situation has built-in conflicts. The Bell telephone companies are expected to help their competitors gain a stronghold in their territories.

The cost of interconnections between existing local telephone companies and competitors, discount levels on resale of local telephone company services and the speediness of access to billing and databases by competitors are all being contested in courts and at the federal government level. There are also jurisdictional disputes over whether discounts on resale of Bell services and prices to interconnect Bell facilities to competitors should be regulated by state utility commissions or by the FCC. Prior to passage of the Telecommunications Act, the availability and terms of local competition were decided at the state level by state public utility commissions. The Act now mandates that the FCC, with the Justice Department, approve these agreements and pricing.

Although provisions of the Telecommunications Act of 1996 are tied up in federal and state courts, competition is moving ahead. In particular, large organizations, firms with a heavy percentage of local calling and/or multiple sites within the same state are being offered services by new local exchange providers. These multisite organizations and organizations with heavy volumes of local calling include hospitals, colleges and universities, government agencies, private schools and headquarters offices of conglomerates. Even smaller businesses with more than one office within a local area have options for local services from multiple providers of data and voice services. The impact of increased local competition will take longer to reach small businesses, homes and people that work from their homes. In particular, these smaller sites need higher speed telephone lines for access to the Internet and work-at-home applications.

Finally, the availability of local number portability promotes local competition. Businesses, non-profits and residential users care strongly about keeping their telephone number if they change to competitive local exchange carriers. Local number portability, the ability to keep a telephone number when changing telephone companies, is mandated by The Telecommunications Act of 1996. Local Bell telephone companies are required to provide local telephone number portability by December of 1998 in the 100 largest metropolitan areas. Wireless local number portability is mandated by June 30, 1999.

Competition for interstate long distance services and on-site telephone systems began in the 1960s and was given a boost by the January 1, 1984 divestiture, which was negotiated in the early 1980s (see Table 4.1 for an overview of regulatory highlights). In 1982 when the Justice Department mandated that AT&T divest itself from the local Bell telephone companies, the Justice Department took the stand that interstate, out-of-local-region long distance, was a competitive service, but that local telephone service was a natural monopoly. The view that local telephone service is a natural monopoly eroded gradually during the 1990s. During this time, competition was gaining a foothold in metropolitan areas, but was not generally available in more rural areas. Technological advances played a role in furthering competition for local calling services.

Specifically, the growing availability of fiber optic cabling and wireless technologies increased the viability of competition in the local loop. Instead of laying miles of individual copper for each customer, competitive telephone carriers now supply a few strands of fiber capable of serving the needs of large corporations. Moreover, improvements in signaling technologies brought down the cost of provisioning local service by requiring fewer and less sophisticated switches to route calls. Powerful computers with databases containing customer telephone numbers and telephone service configurations are provisioned in networks. Telephone company switches, which share the databases, therefore do not require a high degree of complexity. These innovations are bringing down the cost of competing for local telephone service.

Table 4.1   Regulatory Highlights

Landmark Acts and Court Rulings	Summary of Acts and Rulings
The Federal Communications Act of 1934	Congress created the Federal Communications Commission and gave it the authority to regulate interstate telephone, radio and telegraph companies.
The 1956 Consent Decree	The Justice Department allowed AT&T to keep its monopoly, but restricted it to common carrier functions. The Consent Decree mandated that any patents developed by Bell Labs, then AT&T, be licensed to all applicants requesting them. This led to microwave technology's availability to MCI and the ability of competitive carriers to build long distance networks.

Table 4.1 Regulatory Highlights *(Continued)*

Landmark Acts and Court Rulings	Summary of Acts and Rulings
The 1969 MCI Case	The Federal Communications Commission ruled that devices other than AT&T devices could be connected to the network providing the network was not harmed. This decision opened the CPE market to AT&T rivals such as Rolm and Executone.
The 1982 to 1983 Modified Final Judgment	The Justice Department, in agreement with AT&T and with approval by Judge Harold H. Greene, agreed to a settlement that: • Divested the then 22 Bell Operating Companies (BOCs) from AT&T. • Prohibited BOCs from inter-LATA long distance, sale of CPE and manufacturing. • Mandated that the local exchange companies provide equal access (dial 1) from end-users to all interexchange carriers.
The 1984 Divestiture	The terms spelled out in the Modified Final Judgment were implemented on January 1, 1984. The 22 Bell telephone companies were merged into seven Regional Bell Operating Companies, RBOCs. The RBOCs were allowed to sell local and toll calling within the 197 defined local or LATA areas. They also retained the yellow pages. AT&T kept manufacturing, inter-LATA and international toll calling.
The Telecommunications Act of 1996	Decreed that cable companies, electric utilities, broadcasters, interexchange carriers and competitive access providers could sell local and local toll calling. • Allowed local competitors interconnection to and resale of local telephone companies' facilities. • Set fees for interconnection services at the LECs', local exchange carriers', costs plus a reasonable profit. • Set fees for resale at local exchange carriers' costs. • Allowed Bell companies to immediately provide out-of-region long distance. • Allowed Bell companies to provide inter-LATA toll calling and manufacturing in their regions under FCC approval or by February of 1999, whichever is earlier. • Dictated that FCC approval depends on the incumbent LEC's[*] meeting conditions of a 14-point checklist of opening its regions for competition.

[*] The term incumbent LEC, local exchange carrier, refers to the Bell Operating Companies.

Lowering the barriers for entry into the local telephone service market increases the availability of high-speed local telephone services. An array of sophisticated features are obtainable to customers in association with their interstate long distance services. In particular, interexchange carriers such as AT&T, MCI and Sprint have long sold sophisticated 800 services that route calls based on the day of the week, the location of the caller and the number of calls routed to each site. These services are not universally available from local telephone companies. Moreover, interexchange carriers are "first on the block" with efficient data services such as frame relay, ATM and virtual private data networks. Competition is the key reason for differences in the availability of telecommunications offerings from local and interstate carriers. Interexchange carriers have an incentive, in the form of competition, to add innovative telecommunications services. The framers of the Telecommunications Act of 1996, want to advance these innovations and improvements in local calling services by fostering competition. The challenge is to make enhancements affordable to small customers.

Factors Leading to Passage of the Telecommunications Act of 1996

Factors important in pushing passage of the Telecommunications Act of 1996 included:

- Improvements in fiber optic and signaling technologies.
- The Bell telephone companies' lobbying efforts for permission to enter interstate long distance and manufacturing.
- The interexchange carrier's push for entry into the local calling market.
- Viability of wireless services as a substitute for wired local telephone lines.
- Interstate long distance companies' sales of local toll calling to large customers.
- A desire in Congress to allow competition for local telecommunications services uniformly in all of the states.
- A government effort to make access to high-capacity telecommunications services "universal" and affordable.
- Increasing demand by customers for telecommunications services to carry data, image, color and video.

Regional Bell Companies' Desire to Expand Their Offerings

While the seven Regional Bell Operating Companies (RBOCs) wanted to retain their local monopolies as much as possible, they also wanted to expand their range of offerings. They were being hit by competition for their most lucrative customers—big businesses in downtown, metropolitan areas. Moreover, Bell companies were cash-rich. They had entered other businesses with uneven success that required expertise beyond their core capabilities. For example, NYNEX tried and dropped several ventures in value-added computer service companies.

Interexchange Carriers', Utilities' and Cable Companies' Desires to Enter New Markets

In spite of competition, demand for calling services is growing so quickly that even with loss of market share, Bell telephone companies' local and local toll revenue continues to climb. According to its annual report, Ameritech toll revenue increased from $5,586 million in 1995 to $6,068 in 1996. Cable companies, electric utilities, competitive access providers (CAPs), interexchange carriers, resellers and agents all want a piece of this lucrative pie.

Interexchange carriers are eager to enter new markets. AT&T, for example, has seen its market share drop from 90.3% in 1984 to 54% in 1997. Other carriers are battling competition from carriers and resellers. Cable companies are also looking at the long distance market as an avenue for growth. They feel they can leverage their existing investment in fiber and coaxial cabling for entry into local and toll calling.

A quote from the March 10, 1997 issue of *Time Magazine* by Thomas May, the CEO of Boston Edison, typifies many potential local exchange carriers' thoughts: "The utility industry has been growing at 1% a year for the past decade, whereas the telecommunications industry has shot up 30% a year over the same period."

Power companies are actively exploring opportunities in telecommunications. They are attracted by the growth potential, and, like cable companies, wish to leverage their investments in outside cabling.

Legislative Efforts

A speech delivered by Vice President Al Gore on January 11, 1994, prior to the passage of the Telecommunications Act, helped raise the country's consciousness about the need for high-speed telecommunications services. The address, delivered to the Academy of Television Arts and Sciences, proposed national telecommunications reform. The following is a quote from the office of the Vice President's press release about the speech. The press release outlines five main telecommunications principles of the Clinton Administration:

- Encourage private investment.
- Provide and protect competition.
- Provide open access to the network.
- Avoid creating information "haves and have nots."
- Encourage flexible and responsive government action.

This speech became known in the industry as the "information superhighway" speech. The quote with that phrase follows:

Today we have a dream for a different kind of superhighway—an information superhighway that can save lives, create jobs and give every American young and old, the
chance for the best education available to anyone, anywhere.

Prior to passage of the 1996 Act, other bills were initiated in the U.S. House and Senate and failed. In 1993–94, the Markey/Fields HR 3636 passed the House by 423 to 4 votes, but was not approved by the Senate. Other legislative efforts include Brooks/Dingell's H.R. 3626 and Danforth/Inouye's S. 1086.

The final version of the Telecommunications Act was agreed to in December of 1995 by House and Senate conferees. However, it was held up by Republican Senate Majority Leader, Robert Dole, who objected to provisions giving TV stations free spectrum earmarked for digital television. It was passed in February of 1996 when congressional leaders agreed not to distribute licenses for the spectrum pending further Congressional decisions.

Increasing Demand for High-Speed Telecommunications Services

The Federal government felt that increasing the amount of competition in the local loop would increase the availability of high-bandwidth services to local businesses and residential customers. In particular, applications such as downloading radio, television and video on demand from the Internet were envisioned, as was the transmission of medical images in the form of advance radiology images such as MRI results. This additional quote from Gore's "information superhighway" speech illustrates these thoughts:

> Our current information industries—cable, local telephone, long distance telephone, television, film, computers, and others—seem headed for a Big Crunch/Big Bang of their own. The space between these diverse functions is rapidly shrinking—between computers and television, for example, or interactive communication and video….That's the future. It's easy to see where we need to go. It's hard to see how to get there.

Technological Capabilities to Provide High-Speed Services at Low Costs

Improvements in high-speed computers and fiber optic technologies were bringing down the cost of building telecommunications facilities. This trend started in the 1960s with microwave and continued with T-1, T-3 and fiber optics. It takes less labor and fewer outside cables to transport more data, faster. In addition, the cost and size of switches are decreasing while their power, speed and capabilities are increasing.

The Viability of Wireless Services for Local Exchange Service

In 1980, every local telephone company was granted the right to provide advanced mobile services, including cellular service in half the spectrum allocated for cellular service in each local area. The other half was auctioned to the highest bidder. In 1993, President Clinton mandated that auctions be held for frequencies within the Personal Communications Services, PCS, airwaves. Frequencies in each metropolitan area would be awarded to the five highest bidders with some awards given to minority and small businesses. This was seen, in addition to adding to the federal coffers, as a way to open up competition to existing cellular service providers. It was thought that this would drive down the cost of wireless services to the point where they would be a viable form of competition to local, wire-based exchange service.

> ### LATAs Defined
>
> LATAs, local access and transport areas, were created by the Justice Department in 1984 to define the contiguous geographic areas in which local Bell telephone companies could provide local and long distance services. Calls *between* the 197 LATAs are carried by interexchange carriers. Calls *within* LATAs are carried by local exchange carriers, or co-carriers. States with small populations such as Maine, Alaska and Wyoming are made up of one LATA. Thus, US West, the Bell company serving Wyoming, is allowed to provide long distance to all sites within Wyoming. Massachusetts has two LATAs, one for the eastern part of Massachusetts and one for the western section. Calls within the eastern LATA can be transported by Bell Atlantic. However, calls between the eastern and western LATAs must be carried by interexchange carriers. California has eleven LATAs and New York State has eight LATAs.

The Desire for a Uniform National Policy on Local Competition

Competition for local telephone service proceeded at an uneven rate in the late 1980s to 1990s. Each state public utility took varying stands toward allowing intra-LATA toll competition and approving rates for competing carriers. The northern states allowed more competition than the southern states. Members of the Congress and Senate felt that it was important to have a uniform policy and guidelines for implementation of connections between local Bell companies and competitors.

The Telecommunications Act of 1996 is the result of many factors. Demand for passage of the Act had been building up in Congress for many years. The Telecommunications Act of 1996 opened competitive choices for local calling to residential, small businesses and large organizations. In fact, competitive services that bypassed the local telephone companies had been available to large customers for many years. The Act provided an impetus to a process that was already in motion.

Background on the Telecommunications Act of 1996

Ways Interexchange Carriers and Competitive Access Providers Provisioned Local Telephone Service Prior to 1996

Prior to 1996, competitive access providers, CAPs, such as Metropolitan Fiber Systems and Teleport Communications Group, as well as long distance giants AT&T, Sprint and MCI were taking steps to provide local telephone service. However, market penetration of these services was uneven. Local telecommunications was, for the most part, regulated by state public utility commissions. Each state allowed the local telephone companies different levels of monopolistic control of local and intra-LATA toll calling. As state utility commissions opened intra-LATA toll calling, interexchange carriers and CAPs heavily promoted ways customers could use other than the Bell Operating Companies for local toll calling without the benefit of equal access. A great deal of emphasis was placed on California because of its 25% share of the local calling marketplace.

Fiber Optic Lines in Local Areas

Other companies, in addition to the local telephone companies, had miles of fiber optic and coaxial cabling in place. CAPs and MCI started building their own local fiber routes in high-traffic, metropolitan areas. What started as a way to bypass local telephone companies in accessing interexchange carriers' service became the basis for new networks for local calling. Once fiber was in place for carrying calls from commercial customers to interexchange carriers, these lines could also be used to carry data traffic between company sites for purposes such as transporting email and inventory status within metropolitan areas. They could also be used to connect organizations directly to the Internet. In addition to CAPs and MCI, cable telephone companies also had wiring in place. The cable company wires ran down streets in residential as well as in commercial areas. In fact, CAPs often used cable companies to run fiber for them in suburban areas.

Competitors' Local Telephone Switches

Competitive access providers installed their own telephone switches for routing telephone calls. Once the fiber lines were in place, it made sense for many of the CAPs to add their own central office switches as well. Prior to the Telecommunications Act of 1996, CAPs were selling outgoing local, toll calling and Centrex services (see Chapter 2). In New York City in particular, NYNEX was losing many exchange lines associated with downtown Manhattan companies to Teleport and MFS.

Co-located Competitive Central Office Switches with Bell Switches

CAPs started co-locating their own switches next to telephone company switches. On September 17, 1992, the FCC allowed the local Bell telephone companies to open up competition by permitting co-location and virtual co-location. With co-location, local competitors install their own central office switch in the same building as the local telephone company's switch. This gives competitors, usually CAPs, access to Bell telephone company lines for egress and/or termination of telephone calls from the CAPs' customers. In states such as New York, Illinois and California the state regulatory agencies set rules such that co-location took place. This made it easier to pass the Telecommunications Act because the RBOCs were ready to compromise on local calling control to gain access to interstate calling. Local competition was a *fait accompli* in certain places. As the telephone companies lost access lines to competitors, they became more eager to enter new markets such as manufacturing and long distance to make up for losing business in local markets. They were more willing to compromise on opening their own markets to competition in return for entry into intra-LATA long distance. In large metropolitan areas such as New York City, competition was a reality. This was not the case in the south where competition for local calling was not making headway.

What Is a CIC CODE?

A CIC code is a four-digit code used for billing by local telephone companies to exchange carriers and by customers to reach carriers. A CIC code, pronounced kick code, is short for carrier identification code. A CIC code is comprised of three or four digits which are assigned for the most part to carriers. The CIC code, carrier identification code, is inserted after the digits 10 or 101 to become a CAC code, carrier access code.

Each carrier is assigned a three- or four-digit CIC code. For example, AT&T's CIC code is 288, MCI's is 222 and Sprint's CIC code is 333. Users dialing 10XXX, before they dial a telephone number, have their calls routed on whomever's network the CIC code is assigned. For example, someone dialing 10222 will have his or her call routed on MCI's network.

AT&T had a huge marketing advantage because their CIC code, 288, spelled ATT on the telephone dial pad. They were able to use the slogan "Call 10-ATT" to encourage customers to dial 10288 when making local toll calls.

Because the number of available three-digit CIC codes is low, 999, the format has changed from three-digit CIC codes to four digits. The four-digit codes are now being phased in. MCI's CIC code will change from 222 to 0222. CAC codes are changing from 10 to 101. The new CAC code to reach MCI will be 1010222. A transition period during which both five-digit and seven-digit CACs will work starting January 1, 1998. Only the four-digit CICs and corresponding seven-digit CACs may be used starting July 1,1998.

How Callers Access Competitive Local Carriers Without True Local Competition

Dial Around—10XXX

By mid-April of 1994, all but six states allowed intra-LATA toll calling using CIC codes. During the 1990s, long distance resellers and interexchange carriers promoted "dial around" to mitigate the local telephone companies' lock on local calling. They engaged in heavy promotions of the use of 10XXX codes for local toll calling. In particular, AT&T, whose market share had fallen in the first quarter of 1996, according to the FCC, to 54.7% of industry toll revenues of the interstate long distance market following the 1984 divestiture, hoped to gain market share by selling local toll services. In an effort to gain intra-LATA market share, companies started promoting their carrier identification codes for customers to use to bypass the local telephone company's services. Before the Telecommunications Act of 1996 mandated choice in dial 1 local toll calling, residential and small business customers were routed over the local telephone company's lines for intra-LATA toll calls. To get around this limitation, carriers and resellers conducted media blitzes and mail campaigns promoting their CIC codes.

Now that CAC codes with the new format, 101XXXX, have been mandated to take effect by January 1, 1998, advertising campaigns by companies such as Dime Line, with a CIC code of 10-811, and Dial & Save, with 10-457, will have to change. Their new CAC codes, respectively will be 1010811 and 1010457. Carriers currently receiving new CIC codes are receiving four-digit codes in the 5000 and 6000 ranges. Their codes will be 1015XXX and 1016XXX.

Automatic Dialers with 10XXX

To increase sales of local toll traffic, interexchange carriers supplied customers with automatic dialers. These dialers were preprogrammed with carrier access codes to route intra-LATA, local toll calls to the interexchange carrier such as Sprint, AT&T, LDDS and MCI. Carriers also urged customers with more advanced PBXs to program their PBXs such that the PBX, when it recognized an outgoing call as a local toll call, inserted the 10XXX code and routed the call to the customer's interexchange carrier rather than the local telephone company.

What Is Local Toll Calling?

Explaining the concept of local toll calling to customers is a major marketing issue. This is particularly true in states with many LATAs where local telephone companies are not allowed to carry calls to many locations within the same state. Customers don't understand who can carry their calls between these in-state area codes. The concept of a local Bell being allowed to transport toll calls within a local area is confusing. For example, Bell Atlantic carries local calls within the 908 area code. It also carries local toll calls within the 908 area code. However, calls to other area codes within New Jersey, for example to 609, must be handed off to interexchange carriers.

This is a particularly sticky issue in the "dial around" market. Companies in the "dial around" market such as AT&T, Budget Call, Talk Cents and Dial & Save promote their CIC codes, 10XXX, to end-users as a way to sell long distance. To lessen consumer confusion, AT&T, on one mailing to Cleveland, Ohio residents, included a map indicating the area covered by local toll calls.

Bypass Direct Lines from Customers to Interexchange Carriers

By the early 1990s, most large customers with over $5,000 in monthly inter-LATA toll calling had direct T-1 connections from their PBX to their long distance carrier. These T-1 connections enabled 24 voice or data calls to bypass local telephone company access fees on outgoing and incoming toll calls. Many customers with these connections to interexchange carriers programmed their telephone systems to use these direct T-1 lines to carry in-state toll calls via their interexchange company lines (as shown in Figure 4.1). This took another chunk of business away from the local telephone companies. These in-state toll calls were programmed to go out the long distance company's line directly to the interexchange carrier's POP (switch). The carrier's switch then sent the call to the local telephone company office to be routed to its in-state destination. The interexchange carrier kept the outgoing revenue and paid the local telephone company for terminating the call.

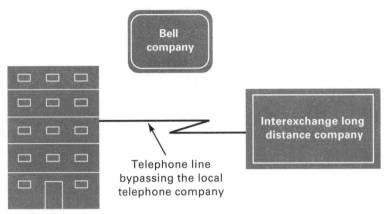

Figure 4.1   Bypassing of the local phone company to save money on toll calls.

The Telecommunications Act of 1996

A key proviso of the Telecommunications Act of 1996 is that it takes away from each state the ability to approve competition in local telecommunications. It lays down a time frame and a method whereby competition will be opened to a variety of vendors. It also outlines a procedure where local telephone companies can expand their operations into manufacturing and inter-LATA, in-region and out-of-region, telecommunications. State utilities are prohibited from denying entry of any qualified entity into interstate or intrastate telecommunications service. It preempts the states' authorities in these regards.

The Act redefines the responsibilities of the state public utility commissions versus those of the Federal Communications Commission. Essentially, it is up to the states to approve rates for local calling and resale and interconnection of Bell services to competitors. However, resale rates charged by the LECs cannot be above their own costs.

Major Features of the Act

Interconnection—Making Bell Resources Available to Competitors

Incumbent local exchange carriers must supply the following to all entities (competitive local exchange carriers) approved by state public utilities as carriers:

- **Resale:** All telecommunications services it supplies to its retail customers should be available to competitive carriers.

    The wholesale price for resale of call transport and termination of calls must be at cost. No provisions of profit for incumbent LECs are built into this ruling.

    Note that the interconnection, as opposed to resale, of unbundled elements of LEC facilities (see below) does include a provision for profits.

- *Number portability:* Telephone number portability to be spelled out by the FCC such that customers can keep their telephone number if they change vendors.

- *Dialing parity:* Equal access such that customers can access their vendors without dialing extra digits. Access to operator services, directory assistance and directory listings with no unreasonable dialing delays.

- *Access to rights-of-way, including poles:* Utilities must provide non-discriminatory rates to cable TV companies and telecommunications carriers other than incumbent LECs. They must also allow non-discriminatory access at equitable rates to their ducts and rights-of-way. If a utility uses its own poles, ducts and rights-of-way to provide telecommunications services, it must charge its own entity the same rate it charges other carriers.

- *Reciprocal compensation:* Telecommunications carriers must establish reciprocal compensation arrangements for the transport and termination of each others' calls.

- *Unbundled access:* Incumbent local exchange carriers are required to provide unbundled network elements such that requesting carriers can combine these elements to provide telecommunications services. A network element is defined as a facility used to provide telecommunications services. Examples of network elements are transmission between central offices, subscriber numbers, information sufficient for billing and collection, systems used in the transmission or routing of calls, signaling and network database access for "smart" features such as caller ID, and operator and directory assistance. Essentially, competitors should be able to buy only the pieces of the Bells' network services that they need to provide local telephone services. For example, an end-user, buying local service on a resale basis from a company such as Cable & Wireless, may have a BellSouth repair person come to his or her home to repair the line if the line was resold to him or her by Cable & Wireless.

- *Co-location:* The incumbent carrier should allow physical co-location of competitive equipment at the incumbent's premise for access to unbundled network features.

Interconnection Agreements—A Timetable

Upon receiving a request for interconnection, an incumbent local carrier may negotiate and enter into a binding agreement with the requesting carrier. These agreements are subject to approval by the state public utility commission. Either party in the negotiations may request that the state utility mediate the agreement. Either party may request arbitration from the state commission from the 135th to the 160th day after the request for negotiation. The state commission shall then be responsible for establishing the rate and schedule for interconnection. According to the Act, the cost for these network elements shall be *based on the cost of providing the interconnection or network element and may include a reasonable profit.* The Act includes a timetable during which the state commissions must rule on interconnection agreements. They can approve or reject selective pieces of the agreements.

Terms that Apply to All Exchange Carriers

The following terms apply to *all* local exchange carriers, including Regional Bell Operating Companies, competitive exchange carriers and independent operating companies:

- Resale of their services.
- Number portability.
- Dialing parity (equal access).
- Access to poles, ducts and rights-of-way.
- Reciprocal compensation for resale of transporting other carriers' calls.

The co-location specifications, interconnection agreements and resale at wholesale rates apply only to the incumbent LECs.

Prohibitions of Barriers to Market Entry— State Agencies Must Allow All Carriers to Compete

The Telecommunications Act disallows any state agency from prohibiting any qualified entity from providing interstate or intrastate telecommunications service. The state commissions are allowed to preserve public safety and welfare and universal service. They are also allowed to manage access, in a neutral way, to rights-of-way.

Some people in the industry, particularly incumbent telephone companies, call new entrants in local telephone service, CLECs, competitive local exchange carriers. The Telecommunications Act of 1996 simply refers to non-incumbent, often non-Bell companies, as exchange carriers.

Universal Service—Affordability and Availability

People in rural, insular and high-cost-to-reach areas, as well as poor consumers, should have access to advanced and interexchange telecommunications services at reasonably comparable rates charged for similar services in urban areas. This provision applies to schools, health care providers and libraries.

Every interstate carrier must contribute equitably to a fund for universal service. State commissions may also create funds for universal service.

Prohibitions Against Slamming—Unauthorized PIC Changes

The Act expressly forbids any carrier from submitting or changing a subscriber's telephone exchange provider without authorization by the subscriber. State commissions are allowed to enforce these procedures in regard to intrastate services. A carrier that "slams" a customer, makes an unauthorized change, is liable to the previous carrier for all fees paid by the "slammed" customer.

Bell Company Entry into inter-LATA Services— State-by-State Approval by the FCC

The Bells are not allowed to sell long distance within their regions until the FCC is satisfied that their networks, on a state-by-state basis, are open to competition. However, the Bell Operating Companies, incumbent local exchange carriers, are allowed to enter the local in-region long distance market 36 months after passage of the Telecommunications Act of 1996 if an agreement has not been signed by then. This date is February of 1999. There is a provision whereby the FCC can extend the 1999 date.

The Bell telephone companies are allowed to provide in-region, inter-LATA calling services when they meet a 14-point checklist called a "competitive checklist." Approval for having met the 14-point checklist will be granted by the FCC, not the states.

However, Bell Operating Companies, according to the Act, *may provide out-of-region* inter-LATA services other than toll-free, 800-type services that terminate in-region. They are also prohibited from selling dedicated out-of-region lines that terminate in-region. These are services where one end is out-of-region and the other end is in-region.

Potential of In-region Long Distance for Bell Companies

The following quote from a May 27, 1997 article in the *Wall Street Journal* illustrates the potential market share Bell Operating Companies have of their in-region inter-LATA markets. "GTE in just a year has taken more than 6% of the available long-distance customers in its territories - more than 82% of these AT&T subscribers. *Southern New England Telecommunications Corp.* has taken more than 30% of the long distance market in Connecticut, mostly from AT&T."

GTE and Southern New England Telecommunications Corp. are non-Bell independent telephone companies. The provision barring BOCs from in-region long distance until they meet requirements in the Telecommunications Act of 1996 applies solely to Bell companies. It does not restrict any of the 1300 independent telephone companies from entering the interstate, inter-LATA long distance market.

The states with the largest number of telephone customers are: California, Florida, Illinois, New York and Pennsylvania. Note that New York and Pennsylvania, with the merger of Bell Atlantic with NYNEX, are both in Bell Atlantic territory. None of these states is in US West territory.

FCC approval will be granted on a state-by-state basis for permission of the Bell companies to sell *in-region* long distance. Approval will be based upon agreements reached with competing carriers that include the items on the 14-point checklist. The agreement must be with a *facilities*-based carrier, which is a carrier that uses predominately its own switches and cabling for carrying customers' calls, unless no facilities-based carrier has requested interconnection by ten months after passage of the Act.

The 14-Point "Checklist"

The 14-point checklist is designed to ensure that the BOCs have opened their networks to competition before they are allowed to sell in-region inter-LATA long distance services. All of the following items must be agreed to in agreements with competitors before the Bells are granted approval to sell in-region inter-LATA long distance:

1. Interconnection as stated above.
2. Non-discriminatory access to network elements as stated above.
3. Non-discriminatory access to Bell-owned poles, ducts and rights-of-way.
4. Unbundled local loop transmission from the central office to the customer premise.
5. Unbundled transport from the trunk side of the local switch (trunks are telephone lines that run from one central office switch to another central office switch rather than to an end-user).
6. Unbundled local switching (routing calls) separate from transmission services
7. Nondiscriminatory access to 911, E911, directory assistance and operator call completion.
8. White pages directory listings for competitors' customers.
9. Nondiscriminatory access to telephone numbers by competitors' customers until numbering administration has been given to organizations other than Bell companies.
10. Nondiscriminatory access to databases and signaling required for call routing and completion.
11. Interim telephone number portability via remote call forwarding or direct-inward dialing until new arrangements are complete for full number portability.
12. Nondiscriminatory access to services that allow competitive carriers to supply dialing parity, i.e., dial 1 to access customers' PIC (primary interexchange carriers) for local toll and non-toll calling.
13. Reciprocal compensation arrangements for Bell and competitive carriers to carry each others' calls.
14. Resale of telecommunications services as stated above, at cost, without provision of profit for the Bells.

Manufacturing by Bell Operating Companies— Concurrent with Approval for In-region Long Distance

The Bell companies are allowed to manufacture and sell equipment once they are allowed entry into inter-LATA toll calling. However, they are not allowed to jointly manufacture equipment with a Bell with whom they are not affiliated. They may, however, collaborate with a manufacturer on the design of equipment prior to their approval to manufacture equipment. They are allowed to receive royalties on these efforts.

Separate Subsidiary Requirements for Manufacturing, inter-LATA Information Services and inter-LATA Long Distance

The Bells are required to sell in-region inter-LATA long distance and manufacture equipment and provide inter-LATA information services through separate subsidiaries.

The separate affiliate status is required for three years after approval for in-region long distance for manufacturing and long distance. It stays in effect for four years for inter-LATA information services. The FCC is given the discretion to lengthen the period required to sell these services via separate subsidiaries.

The BOCs are not required to form a separate subsidiary to sell out-of-region long distance and electronic publishing and alarm monitoring services.

Electronic Publishing by Bell Operating Companies

The BOCs are allowed into electronic publishing through a separate subsidiary or in a joint venture less than 50% owned by a Bell company. Electronic publishing is defined as provisioning or disseminating a variety of news, sports and informational material. Basic exchange services cannot be used to disseminate this material. These rules apply until the year 2000.

Broadcast Services—Relaxed Rules

Decisions on the allocation of spectrum for digital television are left to the FCC. Rules on radio and TV station ownership were relaxed. For example, the Act states that the FCC must remove the limit of one radio and one television station an individual organization can own. In addition, it allows one party to own both a broadcast and cable TV system. It also eases the rules on renewal of broadcast licenses.

Cable Services—Deregulation of Rates

The Act allows cable TV rates to be deregulated once effective competition in the serving area exists. Effective competition is defined as the offering of comparable video programming by a local exchange carrier or its affiliate. Premium services will be deregulated by March 31, 1999.

Video Programming Services—A New Source of Revenue for Telephone Companies

Exchange carriers are allowed to offer video programming, but if they do, they will *not* be required to make capacity on their system available to competitors on a nondiscriminatory basis.

The following rules apply to video programming services:

- Cross ownership of more than 10% of cable TV and local exchange carriers is prohibited.

- A telephone company may operate as an "Open Video System," OVS. If a telephone company chooses to operate as an OVS, it is not required to obtain town-by-town local franchises.

- Local authorities may charge fees that are a percentage of their revenue.

- Local exchange carriers that offer video programming are required to set aside channels for other companies' programming.

- They need to obtain FCC certificates to operate.

- The exchange carrier may obtain joint use of the cable TV providers' "drop" wires from the last multiuser terminal (e.g., telephone pole) to the end-user's premise.

Broadcasting Obscenity and Violence

Regulations on pornography on the Internet and indecent material available on the Internet were passed. However, these were found unconstitutional by the Supreme Court on June 12, 1996.

Interconnect and Resale Exemptions for Rural Telephone Companies

The interconnection and resale rules intended to promote competition do not apply to rural telephone companies. Rural telephone companies are defined as those with fewer than 50,000 access lines. In addition, smaller local exchange carriers can apply to their state public utility commission for exemptions to these rules if they feel the rules impose a financial hardship.

According to a page B1 *Wall Street Journal* article of August 19, 1996, these rural companies cover 10% of the U.S. population. This equates to 10% of the population who may not receive the lower prices and advanced services envisioned as resulting from greater competition for local calling. Some of the multimillion dollar companies mentioned in the article as planning to request waivers for interconnection are GTE, Alltel Corporation, Frontier and Cincinnati Bell Telephone.

Competition in the Local Calling Market

There are technical, legal and financial considerations in provisioning local telephone service. A complex system of connections between telephone companies must exist. Understanding how calls are passed between competing local carriers and between local carriers and interexchange carriers is important in comprehending the issues contained in the Telecommunications Act of 1996.

Equal Access—Mandated for Local Calls
by the Telecommunications Act of 1996

Equal access is the ability of a customer to pre-select his or her telephone company so that he or she can use the carrier of his or her choice without dialing a carrier access code, 10XXX or 101XXXX. With equal access, customers' calls are routed to their carrier when they dial a local or local toll call. Prior to the Telecommunications Act, equal access applied only to toll calls carried by interexchange carriers. At the time of divestiture, 1984, the LECs claimed they would not be able to keep local exchange prices low if they had competition.

Equal access requires that competing local exchange carriers have their central offices connected to each other. These connections are required so that a customer on vendor A's network can call a customer on vendor B's network. Billing arrangements and agreements are also needed for vendor A to bill vendor B for terminating vendor A's call. Some new local exchange providers will resell the incumbent Bell company's service rather than build their own fiber runs and central office switch. In these cases, databases in the network must be updated so that the local telephone company knows that telephone number 555-1234 belongs to vendor A and telephone number 555-6666 belongs to vendor B (see Figure 4.2). Customers in local areas with equal access will choose their local telephone company through a process called PIC, primary interexchange carrier.

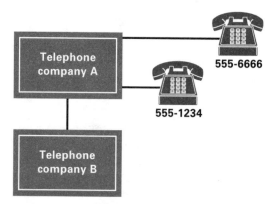

Figure 4.2   Connections between central offices
to achieve equal access.

Primary Interexchange Carriers (PICs)—For Local Calling

Callers select their long distance provider through the "PIC" process, primary inter-exchange carrier. PIC is pronounced pick. Customers who select co-carriers for local and local toll calling do not have to dial the carrier access code, 10XXX, until 1/1/98. This is equal access. It is analogous to the customer's ability to select alternatives such as Frontier Telecommunications, Cable & Wireless, MCI or LDDS for interstate and international calling, without being required to dial extra digits. True competition in the local market can't exist without a process where customers can choose the local telephone company over which their calls will automatically route.

The Telecommunications Act of 1996 mandates that the Bell telephone companies allow competitors' access to telephone companies' transport and switching facilities. Thus, when a customer "PICs" a provider for local or local toll calling, that provider will have their facilities interconnected with the local Bell switches, telephone lines or tele-phone poles as required to route calls between the competitive local exchange provider, CLEC, and Bell companies. The carrier will also have access to Bell features such as voice mail, call waiting, operator services, directory assistance, 911 and repair. The intent of the Act is to enable all customers to have a full range of features and benefits for local service provided by a full array of competitors.

Cream-Skimming and Universal Service—Concentrating on Profitable Markets

Competitive local exchange carriers, CLECs, generally concentrate their initial sales efforts in highly populated metropolitan areas. This concept is known as "cream-skim-ming." Cream is skimmed from lucrative markets such as downtown New York City, where one fiber run has access to thousands of customers in a single skyscraper. This is the reason why competition for local calling started in metropolitan areas such as New York City, Chicago and Los Angeles. There is more potential for profit when an investment in new technology can reach thousands of potential customers in a small area. In a rural area, one fiber run may reach only ten customers.

The economics of telephone service are similar in this respect to the economics of airplane service. Once strands of fiber, microwave service or telephone switches are in place, it costs no more to switch one call than to switch thousands of calls. However, after the break-even point has been reached, the potential for profit is large. If it costs $10,000 to lay a strand of fiber, once the $10,000 is recouped in calling revenue, extra calling reve-nue on that fiber yields high profits because the fiber is paid for and ongoing maintenance expenses are low. This is analogous to flying an airplane. If the fuel and depreciation costs are recouped with 100 passengers, the 101st passenger adds no costs to the flight and is pure profit. For both the airline and telecommunications industries, everyone wants to pack as many customers as possible on each airplane or telephone line.

The desire by the federal government to ensure that telephone service be provided to rural, possibly unprofitable regions is one reason for the regulation of telephone service. The Bell System had, since the early 1900s, promised to supply universal, affordable, basic "dial tone" in exchange for a monopoly. One consideration in opening up local calling areas to competition is how to ensure the continuation of affordable services to poor and rural sections of the country. Telephone providers that only run services to high-profit areas can potentially set their prices lower than Bell Operating Companies who sell telephone services in locations where operating profits are low and expenses are high.

The implementation of universal service may be eroded in rural areas because these telephone companies can apply for exemptions to allowing interconnection services to competitive carriers. The lack of competition has the potential to keep prices high and provide no incentive for implementation of advanced telecommunications offerings in rural locations.

Components of Local Calls

The following must be considered within the context of local calling:

1. *Transport:* The line from a home or business to the central office.
2. *Switching:* The use of the central office to route a call to its destination.
3. *Terminating transport:* The transmission of the call to its end site, or destination.
4. *Signaling:* Signals in the network include telephone number dialed, busy signals, ringing and the diagnostic signal generated by carriers for repair and maintenance of the network.

Transport termination and switching functions are illustrated in Figure 4.3.

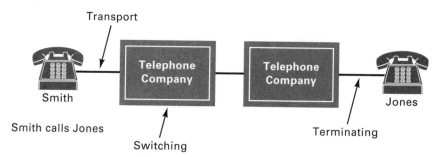

Figure 4.3 Components of local calls, transport, switching and terminating transport.

Other elements of calling include directory assistance, repair reporting, white and yellow pages, 911 and value-added services such as voice mail and caller ID. It takes large capital investments to install and maintain networks. For example, a *Network World* article of October 28, 1996 quoted a figure from Tele.Mac, a telecommunications consulting firm, of $1,800 to $2,000 per each combined copper-coaxial drop. A *Boston Globe* article dated October 1, 1996 stated that according to Yankee Group, fiberoptic drops to multidwelling units cost between $2,500 and $3,500 for every home.

Competitive Local Exchange Carriers and Co-carriers

Competitors to the RBOCs are known as CLECs, competitive local exchange companies, local exchange carriers and co-carriers. Co-carriers have all the rights of local telephone companies, including the ability to sell white page listings, toll-free, e.g., 800 and 888 numbers and 911 services. Potential and actual competitors include: cable TV companies, gas and electric utilities, broadcasters, Bell Operating Companies, independent operating companies, interexchange carriers, agents and resellers.

Three Strategies for Entering the Local Calling Market—Resale, Wireless and Construction of Facilities

Resale. Long distance vendors, agents and resellers resell local telephone company capacity as well as local competitive access provider's capacity. Resale of local services enables new competitors to avoid the cost of building fiber routes or wireless equipment.

Resale of local services to competitors such as AT&T, cable companies and alternative access providers is mandated by the Telecommunications Act of 1996. These potential competitors are hoping for high enough discounts on the price of transport, switching and terminating services to enable them to earn a profit on resale.

The Telecommunications Act of 1996 required that Bell telephone companies *unbundle* the resale of components of local calling. Competitors can choose to buy, for example, only terminating transport in some regions and switching and terminating transport in other regions.

Wireless. Provisioning of wireless services is less labor-intensive than laying copper or fiber cables. As a matter of fact, new telephone services in international locations such as Mexico and Africa are frequently provided via wireless technology (wireless technology is discussed in Chapter 9).

Another factor in the use of wireless is the fact that many carriers already have facilities in place. Consider AT&T's statement in a February 8th, 1996 press release:

> AT&T Chairman, Robert E. Allen...said fixed wireless technology is also an option (as a means to offer local service), adding that 200 million Americans live within the territories where AT&T already has cellular and PCS licenses.

Construction of Fiber Routes. This is appealing to competitive access providers and cable companies who have cabling in place in metropolitan areas. MCI started a subsidiary, MCI Metro in 1994, to develop fiber networks in major metropolitan areas. MCI's local fiber networks are used to transport calls from local customers to the MCI network, as well as for providing local calling.

AT&T has stated that it will use a combination of all three strategies to enter the local telephone market: wireless, resale and its own fiber. For example, a co-carrier may

Anticipation of Local Competition Triggers AT&T's Trivestiture

Prior to 1996, AT&T sold and manufactured long distance services, central office switches, computer systems and business telephone systems. On September 20th, 1995, AT&T announced plans to separate into three publicly-held companies. The trivestiture enabled Lucent Technologies to sell central office equipment to local phone companies without being perceived as a competitor in the arena of local and long distance services.

The break-up resulted in the formation of separate entities to sell telephone hardware (Lucent Technologies), long distance services (AT&T) and computer systems (NCR).

AT&T: annual revenue of $51 billion; 127,000 employees.
- Consumer and business domestic and international calling services, wireless services and on-line services (AT&T Laboratories).

Lucent Technologies: annual revenue of $21 billion; 131,000 employees.
- Central office hardware and software to telephone, cable and wireless service provider companies, integrated circuits, business communications systems and consumer products (Bell Laboratories).

NCR: annual revenue of $8 billion; 38,000 employees.
- Computer systems and services.

choose to construct its own fiber routes in areas where it has many customers, and resell Bell transport to get started initially in regions where it has fewer customers.

FCC Rulings, Legal Challenges and Progress toward Deregulation

Prior to divestiture, the Justice Department and MCI accused AT&T of hindering competition by denying callers access to competitive interexchange carriers. The Bell local telephone companies were divested from AT&T so that AT&T would not be in a position to impede interconnections to the local loop. In contrast to divestiture, the Telecommunications Act of 1996 mandated that the very organizations that competed with new entrants must also supply connections and services for competitors. The local Bell companies are offered the carrot of entrance into new businesses, out-of-region long distance and manufacturing. Nonetheless, conflicts of interest are inherent in pricing for and arranging for resale and access to Bell resources. It is no surprise that issues of pricing for resale and interconnection are being contested in court (see Table 4.2).

Pricing is not the only bone of contention in these cases. Jurisdiction between the FCC and state public utility commission on control of in-state pricing of incumbent exchange carrier services to competitors is a major issue. State public utilities and local incumbent telephone companies have stated that state utilities have the prerogative,

Table 4.2   FCC Rulings and Legal Challenges to the Telecommunications Act of 1996

Date	Decision or Action
June 27, 1996	The FCC spelled out rules on service provider portability. It stated that customers must be able to keep their telephone numbers when they changed carriers. It also stated they must be able to keep "smart" features such as call waiting when they changed carriers.
August 8, 1996	The FCC set rules for calculating the wholesale fees that BOCs could charge competitors for network elements. It also identified seven pieces of the network that must be leased to rivals. The discounts were in the 17% to 25% range. Access fees to wireless companies were reduced by $1 billion annually.
September 12, 1996	Allowed utilities whose lines cross state boundaries into telecommunications.
October 15, 1996	The U.S. Court of Appeals for the Eighth Circuit stayed (denied) the FCC's jurisdiction in setting interconnection and wholesale pricing at the local level. Stayed the FCC's August 8, 1996 ruling above.
October 11, 1996	Justice Clarence Thomas refused to lift the 10/15/96 stay or the Eighth Court of Appeals. Federal regulators asked the ruling to be overturned.
November 11, 1996	The FCC appealed Justice Thomas' ruling. The Supreme Court upheld the Eighth Circuit's October 15, 1996 stay on the FCC's ability to set pricing guidelines.
May 7, 1997	The FCC lowered access fees, the fees interexchange carriers charge to transmit and receive calls from the local networks, by $1.7 billion the first year and $18.5 billion over five years. The FCC also raised end-user line charges by $2.75 for each business line and $1.50 for a second home to pay for Telecommunications Act of 1996 mandated subsidies for schools and libraries.
April 8, 1997	FCC Chairman Reed Hunt postponed a ruling on how to pay for universal service mandated by the Act. Universal service subsidies to poor and rural areas are now funded by access fees carriers pay to connect their networks to local networks and end-user common line charges, EUCLs.
1/1/97 to 6/26/97	Both Ameritech in Michigan and SBC in Oklahoma have applied to the FCC for approval to provide in-region long distance. Ameritech withdrew its filing in January. Ameritech applied a second time and the Justice Department again recommended it not receive approval from the FCC. The FCC denied the SBC application to provide in-region long distance.
October 1997	Eighth Circut Court of Appeals suspended FCC rules on proceedures for interconnection to local networks
December 31, 1997	US District Court excluded October's ban of SBC and US West's entry into long distance. FCC plans to appeal.

granted them by the 1934 Communications Act, of setting resale and wholesale discounts in their states. To date, the Eighth Circuit Court has agreed with this principle. The Supreme Court has yet to take final action on these cases.

Enforcement of provisions and details of implementation of the Act were left, for the most part, to the FCC. The FCC has made rulings on local number portability, access fees and discounts on resale and interconnection. Its rulings on wholesale rates and its rights to set rates have been challenged by the state public utilities, the local telephone companies and independent telephone companies. Challenges to pricing and pricing jurisdictions, who has the right to set prices, are now in court. To date, no incumbent Bell Operating Company has been approved to sell in-region inter-LATA long distance. Both Ameritech and SBC have applied for approvals.

Another major issue, how to fund universal service and discounted access by schools and health organizations, has not been resolved. The FCC was mandated to set forth a plan on funding universal service by May 8, 1997. However, the then chairman of the FCC, Reed Hundt, recommended postponing a plan until the FCC had an industry consensus. Billions of dollars are required to fund these discounts and it is no easy task allocating costs between consumers, businesses and carriers.

Access Fees, Universal Service and Pricing Jurisdiction

State utilities and incumbent local telephone companies are disputing whether the state utility commissions or the FCC has authority over setting pricing standards for interconnection and resale in the federal court system. State utility commissions are approving and acting as arbitrators over interconnection agreements between Bell telephone companies and competitive local exchange carriers. As a matter of fact, many of them are using the 17% to 25% discount off retail suggested by the FCC in its contest ruling of August 8, 1996.

Interconnect Agreements Are Being Set with No Uniform Policy

Each state calculates discounts its own way and leeway is being given differently on non-pricing issues. For example, the state of New York does not require NYNEX to list its competitors' customers in the yellow pages. Moreover, some local telephone companies are more flexible than others in negotiating interconnection agreements. A Bell Operating Company, eager to enter in-region long distance, may be more flexible in negotiations than a BOC with less incentive to offer long distance. For example, US West, with fewer dense pockets of customers, has more to fear from competition than to gain from entering new markets.

New Competitors Are Petitioning State Utility Commissions Because of Unfair Competition

AT&T, which initially offered service in Sacramento, California in January of 1997, pulled out of the market in May of 1997 claiming that Pacific Bell's lack of cooperation hindered AT&T's business. In a complaint reminiscent of MCI's complaints in the 1970s and early 1980s, MCI and AT&T jointly filed a complaint to the California Public Utility Commission on May 27, 1997. They claim that Pacific Bell is not switching customers from Pacific Bell databases to AT&T and MCI databases when customers request them as their local carrier. Pacific Bell has meanwhile accused AT&T and MCI of sabotaging their agreements so that Pacific Bell is not able to offer in-region inter-LATA long distance.

This is reminiscent of events prior to divestiture when AT&T had a lock on access to the local loop for interstate long distance. At that time, MCI sued AT&T over this issue.

Local Telephone Companies Are Speeding Up Deployment of Advanced Services

In an example of new deployment of advanced services, NYNEX publicly announced plans to lay a combined fiber optic and coaxial cabling to the curb capable of eventually reaching one million customers. The new fiber/coax lines will be used for both video on-demand and voice calling in high-volume, lucrative metropolitan areas. The move by NYNEX is part of a broader push by the Baby Bells, in response to competition, to offer residential customers higher speed services.

Not to be outdone by the RBOCs, MCI deployed high-speed SONET and telephone switches in 18 metropolitan areas. SONET transmission is capable of transmitting a variety of customer's video, data and imaging traffic. MCI has received cash for new investments from British Telecom, which purchased a majority ownership.

AT&T, which has invested heavily in fixed wireless technology as a vehicle for providing local telephone service, stated in its April 21, 1997 quarterly report:

> Our continuing investment in new initiated reduced core earnings by approximately 25 cents per share....We began offering local service to consumers in Connecticut, Illinois, and Michigan, and we plan to be in 15 states by the end of the year. We also began offering outbound local service to medium and large-sized businesses in 45 states.

Other companies such as MFS, Brooks Fiber and Teleport are also investing heavily, as are electric utilities. Electric utilities such as Southern California Edison, American Electric Power Company in Columbus, Ohio, KN Energy Inc. in Lakewood, Ohio and Boston Edison are forming joint ventures with telecommunications fiber construction companies. The point of these ventures is to add onto utilities' existing fiber optic cabling and rights-of-way. They also envision that the utilities already have skills in handling billing and customer service issues. Moreover, many electric utility companies jointly own telephone poles with Bell telephone companies. The intention on the part of these utilities is to offer "one-stop shopping" for services such as video on-demand, internet access and voice calling.

Local Access Fees—A Major Source of Bell Company Revenues

Access fees, the fees carriers pay local telephone company for access to customers, amount to billions of dollars annually for the Bell telephone companies. According to Ameritech's 1996 annual report, access fees amounted to 19.6% of their total revenue, $2,938 million. NYNEX, according to their 1995 annual report, received 26.5% of their annual revenue in access fees, $3,557.5 million. The main reason MCI Metro was established by MCI was to avoid paying these access fees to the local telephone companies. MCI wanted to find a more economical way to send calls to and from its customers from its long distance network.

The Yellow Pages—Large Profits, Little Competition

One of the most lucrative business lines for Bell and independent telephone companies are their Yellow Pages. The June 25, 1997 *Wall Street Journal* estimated $11.5 billion in yellow page revenues for 1997. Other organizations have attempted to compete with them, generally with limited success. Rates for listings range from $1400 per month and up for half page ads in major city books. One company with 1,000 employees spends $60,000 annually listing their products. Competition has never been prohibited in the telephone book advertising business. However, the telephone book business continues to be a virtual lock for incumbent telephone companies.

Competition in local telephone services began in the mid-1980s when large business customers bypassed their local telephone companies' central office switches to reach long distance providers. This saved end-users 4 1/2¢ per minute on their toll-free, out-of-state and international calls.

Access fees are intended as a subsidy for local service so that telephone companies can keep rates for residential customers affordable. Residential basic home telephone line fees have always been set at below cost rates. They are subsidized by rates businesses pay and by long distance fees. Rules promulgated by the FCC on May 7, 1997 will lower access fees by $18.5 billion over five years. However, this action shifts costs to residential and business users in the form of higher End User Common Line (EUCL) Charges.

Local Number Portability—Issues and Scheduled Implementation

Local Number Portability—Creating an Equal Playing Field

Local number portability allows subscribers to keep their telephone numbers when they change carriers for incoming "dial tone." Telephone number portability was mandated by the Telecommunications Act of 1996. Without telephone number portability, when a customer changes telephone companies for incoming calling, he or she must change his or her telephone number. The requirement to change telephone numbers when changing telephone carriers for incoming services hinders competition.

Currently, telephone numbers are assigned by the local Bell or independent operating companies in blocks of a thousand to each LEC. For example, each exchange, the three-digit number preceding the last four digits of a telephone number, is assigned to a particular local provider. In downtown Boston, the 223 exchange is served from Bell Atlantic's Bowdoin Street central office. The 330 exchange is served out of Bell Atlantic's Franklin Street central office. When Teleport began selling local exchange service, it was supplied with certain exchanges within the 617 area code, e.g., 563 and 476. In the past, customers that wanted telephone exchange service from Teleport for incoming calls had to change their telephone number to one using an exchange assigned to Teleport's central office.

Achieving number portability requires costly upgrades to older telephone company switches. Interim methods of local number portability do not require these upgrades. These interim methods, described in the following pages, have limitations.

Limitations of Interim Number Portability

The most prevalent form of interim local number portability is call forwarding. With call forwarding, the central office where a telephone number is originally served is programmed to always send calls to the competing carrier's central office.

Use of call forwarding, for interim number portability, has four disadvantages:

1. *Inefficiency*: Call forwarding to achieve telephone number portability uses up to two telephone numbers: one in the Bell telephone company central office in which the telephone number physically resides and one in the competitor's central office. This wastes telephone numbers, ties up extra equipment in the telephone companies' networks and costs extra money.

2. *Unfair competition*: The co-carrier is required to pay an extra charge to the Bell phone company for call forwarding and the use of an extra number in the Bell central office switch. This adds an extra burden, in cost, to potential competitors.

3. *Lack of feature transparency:* Telephone numbers using interim local number portability lose "smart" features including caller ID, call waiting, call return, repeat dialing, call trace and per call blocking. Signaling information required for "smart" features is lost on the forwarding. Not having these features in a carrier's "bag" is a disadvantage to the competition.

4. *Reliability and quality of the call*: Post dial delay, taking longer for calls to be completed, may result from a call taking extra "hops." Reliability, finger-pointing on repair problems and lowered transmission levels, also may result from a call having to travel through multiple trunks and switches.

Four Types of Telephone Number Portability

1. *Service provider portability:* An end-user's ability to keep his or her telephone number when changing carriers.

2. *Location portability:* Keeping a telephone number when moving.

3. *Service portability*: Keeping a telephone number when changing from wireline to wireless or voice to data services.

4. *One number for life portability*: Keeping a telephone number regardless of location or service used.

In complying with the Telecommunications Act of 1996, the FCC has approved service provider portability.

Service Provider Portability: An End-user's Ability to Keep His or Her Telephone Number When Changing Carriers. Costs Are Borne by Carriers.   Service provider portability requires that all telephone numbers within an area be available to all exchange providers. If an end-user decides to use a local utility company who is approved by the state utility commission as a carrier, the customer can retain his or her telephone number and all "smart" features available from the selected carrier. The FCC, based on provisions in the Telecommunications Act of 1996, ordered service provider portability availability in the 100 largest standard metropolitan statistical areas by December 31, 1998.

The method approved by the FCC to accomplish service provider portability is LRN, local routing number. With local routing number, LRN, every central office switch is assigned a ten-digit number. These switch numbers, or LRNs, will reside in network databases. All telephone calls will trigger a "dip" into a database to determine to which central office a call should be routed.

Service provider portability will start in seven cities in March of 1998. Each of these cities is in a different one of the original seven RBOCs. They are:

- Atlanta.

- Chicago.

- Houston.

- Minneapolis.

- New York.

- Los Angeles.

- Philadelphia.

Six months later, 45 more metropolitan areas will be added, and the final 45 will be added December 31, 1998. After December of 1998, local carriers must make portability available six months after receiving a specific request from another carrier. All industry carriers will jointly pay for the establishment and management of the eight LRN databases.

Wireless number portability must be achieved by June of 1999 as follows:

a. Wireless carriers (cellular, PCS and paging companies) must be capable of delivering calls to the wireline "portable" destination by December 31, 1998. Mobile central offices must have a destination switch number, LRN, for look-up capability.

b. By June 30, 1999, customers must be able to change cellular service providers without changing their car phone numbers. All wireless providers must also offer the ability to support roaming within their networks by June 30, 1999.

Location Portability: Keeping a Telephone Number When Moving.   Location portability is the ability of users to keep their telephone number when they physically move within their area. Location portability is not mandated. It is thought that implementation of location portability will be driven by customer demand. The ability for large businesses to keep their telephone numbers when they move is significant.

Service Portability: Keeping a Telephone Number When Changing from Wireline to Wireless or Voice to Data Services.   Service portability enables users to keep their telephone numbers when they change services, i.e., wireline to wireless and POTs to digital data services. Service portability is not mandated. Service provider portability would allow users to keep their telephone numbers when they change to wireless providers for their home telephone service.

One Number for Life Portability: Keeping a Telephone Number Regardless of Location or Service Used. One number for life portability opens up the possibility for out-of-area geographic portability between towns and states as well as between carriers. This is the case with toll-free 800, 888 and soon, 877 calling. These numbers are assigned to customers regardless of their location. No date is set or mandated for one number for life portability.

Summary

The availability of viable competition in local calling services throughout the U.S. requires massive amounts of money and numerous interconnection and resale agreements among carriers. Implementation of the Telecommunications Act of 1996 requires review of interconnect agreements in each state. It also requires judicial review of complex jurisdictional issues between the FCC, state public utilities and local Bell and independent telephone companies. Already, most states are proceeding at different paces on these agreements with varied requirements.

The Act itself is large and arcane. It is 111 single-spaced pages long. Understanding and interpreting the technical and legal components of the Act are difficult. The Act is an outline of the steps and timetables required to be spelled out by the FCC and state commission. Having staffs at these agencies with the continuity and resources to guide the country to full and universal competition is critical to the successful implementation of the Act's requirements.

In addition to legal issues, there are questions of technical resources at the local level. While metropolitan areas of the country have a higher percentage of advanced central offices than rural and poor areas, even metropolitan areas have pockets of old equipment. Frequently, poor neighborhoods in major cities have central offices that impair even standard POTs. For the most part, competition is already spurring investments in cabling, signaling systems and new switches. The first group of consumers to gain from these investments are large corporations, hospitals and universities. Gains for middle-class consumers and small to medium-sized businesses will take longer to achieve.

CHAPTER 5

The Public Network

$\mathbf{T}$he public switched network is considered a strategic asset by the U.S. government. Because of the strategic value of the network, the FCC tracks the reliability of carriers' networks. After major outages in 1991, one of which affected air traffic control in New York, the FCC and Federal Aviation Association (FAA) had talks with AT&T on ways to avoid future outages. The public network is considered so vital that the U.S. Post Office ran AT&T during World War I.

Network services are strategic for individual organizations, as well as for the American government. Corporations and agencies depend on their telephone lines to conduct business. On days when firms lose their telephone service, much of the work of the organization stops. No data or electronic mail can be sent to the Internet or to other corporate sites. Departments such as sales, customer service and purchasing cannot accomplish substantial portions of their work. People who manage and sell telephone services need to understand the concepts of switched and dedicated services. Mistakes in configuring telephone networks result in extra expenses, insufficient capacity and/or maintenance problems.

The goal of this chapter is to review the differences between switched and dedicated services. The main difference between switched and dedicated services is that switched calls are dialable. The connections made by dialing a telephone number are flexible, based on the telephone number dialed. With dedicated services, the links are permanent; a private line from building A in NYC to building B in Washington, DC is a permanent connection. Examples of switched services are home telephone and main business lines.

Switched services have the following characteristics:

- They are dial-up; users dial a telephone number to create a temporary connection.
- People can reach anyone on the public network by dialing a telephone number.

- They are pay-as-you-go; charges are based on the amount of time calls are connected.
- Switched services are used for voice, video, images and data traffic.
- Switched services are used with analog and digital telephone lines.
- When callers hang up, the telephone line and equipment in the network are free to be used by another person or data device.

Dedicated services, also called private lines, are more specialized than switched lines. Organizations use them to save money when they need to transmit large amounts of data or place hours of voice telephone calls to particular sites. Imagine two tin cans and a string between two locations. This is analogous to a private line. Organizations have the use of the string 24-hours-a-day, seven-days-a-week. They pay a flat monthly fee for the use of the line. There are no extra fees no matter how much of the line is used.

Dedicated or private line services have the following characteristics:

- Dedicated, private lines have a fixed configuration.
- Devices and people on the private network can only reach each other.
- Dedicated services are used for voice, video, images and data.
- Users pay a flat monthly fee; there are no per minute charges.
- Dedicated services are used on analog and digital telephone lines.
- Private lines are for the exclusive use of the people or devices connected to them.

A major problem with dedicated, private lines is the time required to manage them. If a firm only has dedicated lines between a small number of locations, maintenance may not be a problem. However, many companies want private lines between multiple sites. They may also need back-up telecommunications services in case the dedicated lines crash. Staff time is required for maintenance and keeping track of equipment inventory associated with dedicated lines. The staff must have the expertise to determine if repairs are needed on the network or on-site equipment. This is problematic for organizations who do not wish to invest money in the staff to manage, design and maintain private facilities. For these reasons, many companies are choosing carrier-managed, value-added network services such as frame relay. (Frame relay is discussed in Chapter 6.)

Network-based computer intelligence has changed the public network from POTs to one capable of delivering advanced features and generating fat profits. Advanced features are made possible by advances in signaling systems in public networks. Large profits are gained from the sale of value-added services such as call forwarding, caller ID and voice mail. According to a *Wall Street Journal* article of September 3, 1996, value-added, smart service profit margins are 70% compared to POTs margins of 10% for Bell and independent telephone companies.

Signaling is the glue that holds the public switched network together. Routing, billing and transferring of calls between carrier networks depend on signaling. Network maintenance information is also carried on signaling systems. Signals include dial tone, dialing strings, busy signals and fast busies in the network. The way signals are transported impacts network efficiency, costs, reliability and introduction of new services. These new services include caller ID, 800, 900 and 500 advanced features such as routing by time of day, integrated services digital network service, ISDN, and voice mail supplied by telephone and cellular carriers.

An issue facing the telecommunications industry is paying for upgrades to local telephone switches for access to all of the network intelligence via Signaling System 7. Signaling System 7 uses high-speed channels in the public network, separate from those on which traffic is carried, to send signals and to access advanced services located in centralized databases. Currently, interexchange carriers and tandem central offices have "hooks" connecting these offices to advanced features in the network. Tandem offices route calls between other central offices and from local to interexchange carriers' networks. However, these hooks, or signaling systems, are not uniformly available at central offices connected to peoples' homes. This is why some people do not have telephone company-based voice mail in their areas. The hooks are also required for local number portability mandated by the Telecommunications Act of 1996. Local number portability allows people to keep their telephone numbers when they change carriers for local telephone service.

Switched Services—Local and Long Distance Calling

Telecommunications is booming. Users in the U.S. are quickly gobbling up new telephone numbers. Consider the rate at which new area codes are used up. The original 144 area codes lasted from 1947 to 1995. From 1995 to mid-1996, 35 new area codes were assigned. Much of this appetite for telecommunications services is taking place on the public switched telephone network.

The public switched telephone network is analogous to a network of major highways originally built by a single organization but added to and expanded by multiple other organizations. Traffic enters and exits these highways (long distance lines) from multiple "ramps" built by still more organizations, e.g., the local telephone companies and competitive access providers.

The original organization that constructed the "highway" system that is the basis of the public switched network is AT&T. Prior to divestiture in 1984, AT&T set standards via its research arm, Bell Laboratories, such that all central office switches and all lines that carried calls met prescribed standards. As a result of these standards, everyone with a telephone can talk to everyone else. Dialing, ringing, routing and telephone numbering are uniform.

Dialed calls are placed using switched services. Any time a telephone number is dialed, a switched service is accessed. The central office, which performs the switching function, routes calls based on the telephone number dialed. This is true for modem, facsimile and voice calls. Carriers and resellers put their own brand names on switched services they sell. MCI sells MCI Vision and MCI One, AT&T offers Uniplan, and WorldCom has WorldOne.

The ITU Definition of Switching

According to the standards organization CCITT, now called the ITU or International Telecommunications Union, "The establishment on demand, of an individual connection from a desired inlet to a desired outlet within a set of inlets and outlets for as long as is required for the transfer of information." In plain English, what goes in (the dialed telephone number) comes out (to the called party) for as long as desired from within the network until one party hangs up. Switched calls carry voice, data, video and graphics.

Attributes of Real-time Switching Services

Addressing—Flexibility

A telephone number is dialed. This is the "address" to which a call is directed. Numbers on a telephone are used to send dual tone multifrequency, DTMF, tones over the network. At the central office these tones or frequencies are decoded to address signals. Area codes are assigned to particular metropolitan areas. Exchanges, the next three digits of a phone number are assigned to a particular central office and the last four digits, the line number is assigned to a specific location. (The assignment of exchanges to particular central offices will change with the advent of local number portability when customers will take their numbers with them if they change to a different local telephone carrier. See Chapter 4, *Local Competition,* for details.)

People who use switched services to transmit data often do so because it provides the flexibility for them to transmit data to multiple locations by dialing different telephone numbers. Switched data services can be used to communicate with multiple locations, not just a limited number of computers or devices. ISDN, switched 56 and plain old telephone services, POTS are all used to carry switched data calls. In particular, people with video conferencing units often use ISDN dial-up, switched services so they can place calls to multiple locations. They are not limited to holding video conferences only with other corporate sites. They can dial into customer, supplier and consultant locations that have compatible video systems.

DTMF: Access to Voice Mail and Computers.   Touch tone, dual tone multifrequency, was introduced by AT&T in 1963 to speed call set up in the central office. Prior to 1963, calls were dialed by pulse signals, rotary dialing. A ten digit call takes 11.3 seconds to dial using rotary pulses. In contrast, a touch tone, ten digit DTMF, call can be dialed in 1 second. Thus, DTMF or touch tone calling adds efficiency to the public network because it ties up the central office switches for a shorter period of time than rotary dialing when call connections are being set up. DTMF uses central office facilities for a shorter time during dialing than pulse dialing (rotary dialing).

DTMF tones are also used to access voice mail, bank accounts and enter orders for home shopping applications from telephones. DTMF is used in voice mail and voice response technology where people send signals to computers via the touch tone pads of their telephones. Once a telephone connection has been established, the network passes any additional DTMF tones entered to the voice mail or voice response system. Thus DTMF tones are available to communicate with computers via voice mail and voice response systems.

DTMF signals are an example of a standard established by the AT&T Bell system so that all callers in the public switched network have a consistent format for addressing calls. The functionality of touch tone dialing has been expanded, as illustrated in the above example, from addressing telephone calls to accessing information in computers and interfacing with voice mail systems.

Pay-as-you-go Postalized Rates

Charges for switched calls are based on the amount of time calls are connected. For example, a ten-minute call costs less than a call lasting an hour. Time-of-day rates may also vary. Peak calling times often cost more than off-peak times. In this way, carriers hope to even out traffic so that they aren't required to build additional facilities to accommodate peak calling patterns.

To save money on long distance charges, companies such as retailers often transmit bulk data files such as inventory status reports during off peak hours, for instance, in the middle of the night.

In the past, the further away a call was transmitted, the more expensive the call was. The cost of long distance service is no longer consistently distance-sensitive. AT&T and Sprint both have flat 10¢ per minute rates and lower-cost plans for domestic calls. This is known as flat rate or postalized pricing. Just as a first class letter costs the same to mail next door or 2000 miles, calls often cost the same whether they carry a conversation to a friend across the state or to a relative across the country.

Once a carrier's high-speed network is in place, it costs the carrier no more to send a call 2000 miles than 400 miles. Capacity is available and carriers want to fill their "pipes".

On-demand

Voice and data calls are initiated by picking up the handset or by instructing a modem to dial a call. The service is available "on-demand". However, callers may find that on peak traffic days, such as Mother's Day, calls will receive a "fast busy". Network capacity, in this case, is not available.

Another example of an instance when services are not available "on-demand" is during major snowstorms. A particularly fierce blizzard in Boston in 1995 resulted in a high number of people working from home and using modems for long stretches of time. This resulted in a strain on NYNEX's network capacity and fast busies on many calls.

Immediate

If capacity is available, service is instantaneous. When someone dials a telephone number, he or she expects the call to be completed immediately. As noted above, extreme conditions can eliminate the immediate capacity expected by users. Natural disasters, unusual demand and human error all impact telephone availability.

The Public Network and National Security

Telephone service is a vital national security and business resource. The FCC and FAA monitor reliability of major carriers' networks. The 1993 earthquake in Los Angeles, software glitches and power disturbances have all interrupted telephone service. Consider the software glitch in AT&T central offices on January 15, 1990 that caused a nine-hour outage. A telephone company power failure on September 17, 1991 in Manhattan also caused havoc.

Carriers build in redundant power sources, remote alarm monitoring, back-up systems, multiple fiber paths to central offices in case of a fiber cut and hurricane-proofing in central offices to ensure continuous telephone service. Customers assume immediate telephone service, which carriers take great efforts to provide.

Analog or Digital—"The Last Mile"

Not all switched services are analog. ISDN is an example of a switched digital service. Basic rate ISDN is a switched digital service with one or two paths that can be ordered from the telephone company. Each path can be used for voice or data. In addition, a third, slower path is used for signaling. These signals are the number dialed, ringing, busy signals, calling party telephone number, on hook and off hook status and alarm signals.

ISDN is requested by schools, telecommuters and businesses for Internet access, access to remote databases, collaborative multimedia computer conferencing and video conferencing. ISDN, while more expensive than POTs provides faster, more error-free transmission services than standard analog lines.

Digital services, which transmit information at higher speeds, are more reliable and contain fewer errors than analog transmissions. Due to work-at-home applications, access to the Internet and medical uses such as x-ray transmissions, more end-users than ever before want high-capacity digital switched services such as ISDN (see Chapter 7). They want to download Internet graphics and large corporate files faster from home.

A major bottleneck of analog services exists in the cabling to residential and small businesses from the telephone company central office. This piece of the network is also known as the "last mile"; it is illustrated in Figure 5.1. It is the leg from the central office's bulk cable to the end-user's location.

In contrast to the "last mile" in residential areas, telephone companies and competitive access providers often lay fiber cables capable of transmitting digital services from their switches to office buildings. The expense of supplying fiber cable to a building with multiple telephones can be spread across many customers. A large office building with many users allows a telephone company (telco) to spread the cost of laying the fiber cable across multiple customers. A fiber cable, on the other hand, which terminates at a single household, must be paid for from the revenue generated by that household.

Local telephone providers, the Bells as well as new entrants in local telephone service, are faced with the challenge of either upgrading existing cabling or constructing new media capable of sending and receiving growing amounts of video, Internet information, entertain-

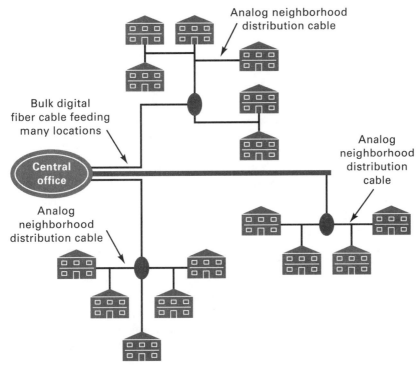

Figure 5.1 *The Last Mile.* Bulk cables, often fiber, from the central office into a neighborhood, use digital transmission, while cables to individual homes often send analog signals. The cable from the telephone pole to the home or business is called a drop. It is the last mile.

ment and multimedia to growing numbers of users. The upgrades and new facilities may take the form of added equipment in the central office, fiber optic cabling, wireless transmissions for local telephone service and upgrades in the copper cables themselves.

Incoming, Outgoing, and Two-way

Incoming 800 and 888 toll-free numbers are popular marketing tools. Companies spend millions of dollars promoting their toll-free, incoming only numbers.

Sales applications using outgoing lines often generate many calls. Many businesses use outbound telemarketing for bill collections and to generate new sales. Specialized computers, predictive dialers, are used to place calls to potential leads for items such as magazine and newspaper subscriptions. These dialers are programmed to rapidly place successive calls. The dialers recognize and hang up when they encounter busy signals, ring-no-answers and answering machines. A live agent's time is not used for dialing, busy signals, answering machines and ring-no-answers. When calls are answered by a live person, the dialer transfers the call to an agent. Agents are used only to speak with live prospects and thus speak with a higher number of prospects each hour.

Predictive dialers are known to bring central offices to their knees because of their high volume of calls. For every agent on a call, the predictive dialer seizes multiple outside lines for dialing, busy signals, ringing tones and listening to answering machines. Central office facilities are tied up for the busy signals, ring-no-answers, answering machine responses and live agent calls.

The total volume of telephone traffic is increasing due to these applications. Use of predictive dialers and sales centers for inbound, toll-free-numbers are two reasons that a carrier such as AT&T can lose market share and still report an increase in the volume of calls carried on its networks. *Organizations recognize that equipment to route and place calls and the cost of calls themselves are cheaper than hiring people to manually dial calls or travel in person to make a sales call.* As cited by Andrew Waite in *The Inbound Telephone Call Center*, McGraw-Hill, 1987, the cost of a business-to-business in-person sales call is $291.10. In contrast, he calculated the cost of a telephone call to sell a magazine subscription at $1.50 for the staffing, telephone equipment and telephone line expenses.

Switched Services for Data—Capacity Issues

The public switched telephone network was designed originally for voice traffic. Voice traffic has different usage patterns than data traffic. For example, voice calls are shorter—an average of three minutes in length. Data calls, on the other hand, tend to be longer. Callers accessing the Internet from homes and businesses often stay on-line for long stretches at a time. According to a 1996 Bellcore traffic study titled "Impacts of Internet Traffic on LEC Networks and Switching Systems" by A. Atai and J. Gordon, calls to the Internet are an average of 20 minutes long with some percentage lasting 12 hours, 24 hours and more.

Residential and business customers use the switched public network for data communications services such as fax transmissions, access to the Internet, video, paging and work-at-home applications. Usage of the public switched network for data has resulted in an enormous demand for longer calls, more frequent calls and additional telephone numbers. In 1995, the structure of area codes was changed to accommodate this demand. The middle digit was changed from only one or zero to any number from zero to nine. This change added capacity for 5,120,000 additional telephone numbers.

Carriers and local telcos originally designed their networks with the assumption that at any given time, not every telephone user would be on a telephone call and that most calls would not tie up the carrier's and local telco's networks for long periods of time. According to *The Business One Irwin Handbook of Telecommunications,* 2nd edition by James Harry Green, central office switches are equipped such that for every four to six telephone numbers, one path for a telephone call exists.

Not equipping offices so that everyone can make calls at the same time saves money on central office equipment. The Atai and Gordon Bellcore study cited above estimated, conservatively, that it would cost each RBOC $35 million per year to upgrade the local telco's networks to keep up with new demand for Internet access in such a way that access to calling would remain at the 1996 level of .05 calls out of every 100 calls blocked.

End-to-end Telephone Service
End-users take for granted the ability to reach people on different long distance networks. For instance, subscribers to MCI assume they can reach AT&T subscribers.
This was not always the case. During the early years of the telecommunications industry, 1893 to 1907, people frequently needed two telephones: one phone for people served by perhaps, The Bell Telephone Company and another telephone to reach people in a town served by an independent telephone company.
Independent and Bell telephone companies did not "speak" to each other. There was no interconnection. AT&T articulated the strategy of end-to-end telephone service in its 1910 annual report. The public switched telephone network grew out of this concept.
The federal government granted AT&T a monopoly on telephone service in return for AT&T's provisioning of end-to-to end telephone service.

Circuit Switching—Not Using the Network Efficiently

The public switched network uses circuit switching to transmit calls. A circuit is a physical path for the transmission of voice, image or data. The ITU, International Telecommunications Union, definition of circuit switching is:

The switching of circuits for the exclusive use of the connection for the duration of a call.

A person or modem-type device dials a call and the network sets up a path between the person initiating the call and the person called. Importantly, the path or channel (the circuit) is available *exclusively* for the duration of the call. The path is not shared. Natural pauses in conversation and data transmission cannot be used for other voice or data calls. Capacity is saved in the network for the entire duration of the transmission. When the call is ended, the circuit is released and the path becomes available for another phone call. This exclusivity causes wasteful utilization of network capacity.

Newer telecommunications services do not have this limitation. For example, ATM is used in portions of the Internet. With ATM, transmissions from multiple voice and data sources share the same path. Pauses in one conversation are filled by data from other sources. Network capacity is not saved for the exclusive use of devices when they are idle. (ATM is covered in Chapter 6.)

Store and Forward—Message Switching

The storage and transfer of messages such as voice mail, data information and facsimile during off-hours is known as message switching or store and forward switching. With message switching, stored messages are transferred at off-peak times to minimize network idle time and network overload during busy times. Store and forward switching does not require sender and receiver to both be available at the time of transmission. Moreover, the network can hold the message and retry multiple times until the receiving equipment is available. For example, in the case of facsimile, if a recipient's fax machine is busy, the sender will retry sending the fax message a preprogrammed number of times at timed intervals.

Not all applications require the real-time, immediate transmission of circuit switching. For example, an organization may send, via facsimile, a price change to 100 sales offices. These price changes are stored in computers and forwarded at a later time. They are often transmitted after hours at a lower cost. This is a good example of a message store and forward application. Customers can program their own fax machines for transmittal at a later hour, or they can use their long distance vendor's store and forward service. When using a long distance vendor, the customer sends the fax once to the carrier who sends the message to each name on the list of recipients. For example, one consultant sends out hundreds of facsimile transmissions to potential sales leads each month. He stores telephone numbers and sales letters in his computer and programs his fax/modem to fax the letters in the middle of the night when long distance rates are lower.

Dedicated Services

Overview of Dedicated Services

Dedicated services, also known as private lines, are analogous to having two tin cans and a string between sites. The "string" and "tin cans" are for the exclusive use of the organization that leases them. The "string" portion is the medium over which the voice or data transmission is sent. The medium is generally copper, microwave or fiber optics. The tin cans refer to equipment owned by the customer such as a telephone, modem, cable modem or terminal adapter. The "tin cans" are the devices which allow sites to transmit computer-generated or voice signals between locations.

Some organizations have such a high volume of voice, video or data calls between locations that they opt to use dedicated, private lines to connect sites together. These dedicated, private lines are available for the exclusive use of the customer that leases them from a carrier or local telephone company. A retail chain in the Boston area has sufficient data communications traffic between its headquarters, warehouses and stores that it has 90 dedicated links connecting end-sites to headquarters for the purpose of inventory status and updates on store sales. This is the chain's own private network.

These dedicated links may be cheaper than paying by the minute for multiple calls to the same sites. Private, dedicated links cost a flat fee per month. Costs are not based on the amount of voice or data transmitted or the amount of time the dedicated links are in use during the month. Another factor in the decision to use dedicated services is the desire for secure transmissions. Some firms believe that the public switched network is too public or open to hacking for applications such as funds transfer.

Large firms may have private networks made up of multiple "strings and tin cans" connecting sites together. These are private networks, available only to the particular company that leases or builds them. Private networks are, as the name implies, private, and they cannot be accessed by outsiders simply dialing into them. The example above of a retail chain with 90 dedicated links is an example of a private network. This is also a wide area network or WAN. The dedicated links connect sites that are located outside of the immediate building or campus area.

Organizations with private networks lease these dedicated lines to interconnect their various sites. Examples of private lines are DATAPHONE Digital Service (DDS), Accunet Spectrum of Digital Services from AT&T and Clearline from Sprint. All local telcos also sell private lines.

Attributes of Dedicated, Private Lines

Fixed Monthly Fees

End-users pay a fixed monthly fee, such as $400 for a line from New York City to Washington, DC. This contrasts to switched long distance services where users are charged per minute fees. With fixed monthly fees, organizations can use dedicated lines around the clock without incurring extra charges. The fee is the same whether the line is used two hours per day or 24 hours per day.

Fixed Routes

Dedicated lines are not flexible. Calls can only be sent between the fixed points to which the lines are connected. Having only the points on the fixed route to which calls can be sent may present a problem in some situations.

Consider room-type video conferencing equipment. Firms may purchase two video systems, one in New York City and one in California for conferencing between these two company sites over a dedicated line. As often happens with video systems, the firm may at some point wish to hold a conference with a customer or vendor. If the equipment is connected only to the private line running between two fixed points, the flexibility of holding a conference with multiple vendors or customers is not available.

Exclusive Use

Dedicated circuits are not shared. They are for the sole use of the customer who owns or leases them. In the public switched telephone network, anyone can call multiple other locations. Not so with dedicated lines. Dedicated, private lines are put into place so that voice or data can be sent exclusively between the points on the private lines. Many organizations own private switches with the intelligence to route calls over their own dedicated lines. However, if the dedicated lines are busy when an additional request is made to place a call, the private switch will send the call over the public switched network. This enables firms to pack a high volume of traffic on their dedicated lines, but have the flexibility to route overflow, peak traffic on the public network.

24-hour-per-day Availability

Dedicated services are available around-the-clock. This is cost-effective for companies that use the dedicated lines for voice, video and email during the day and bulk data transmissions such as transmissions of sales figures after hours. To illustrate, a corporation may use dedicated lines that cost $5,000 per month for video and voice calls between two locations during business hours. After hours, they use the line for data communications. There is no fee for data transmitted after hours; it travels "free" during non-business times.

Voice, Video and Data

Dedicated lines are suitable for transmission of both voice and data. Voice and data can share the same dedicated services or they can use completely different dedicated lines. Firms often lease T-1 lines which have 24 channels to tie locations together. They can use, for example, twelve of the paths for voice and twelve for data or video. See Figure 5.2 for an example of sharing an access line for noise and data.

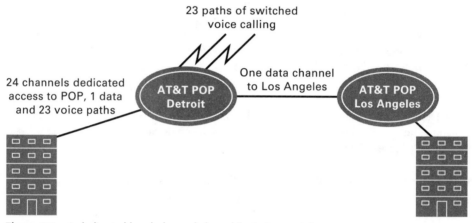

Figure 5.2   A dedicated local channel shared for voice and data.

Dedicated links are often used for shared transmission of voice and data directly from a customer premise to an interexchange long distance provider. This enables customers to save money on the portion of the dedicated line between their premises and their long distance provider. The cost of the dedicated line between the customer and long distance company is shared between voice and data.

The voice and data share the private T-1 from the customer premise to the long distance provider's switch. Once the calls arrive at the carrier's switch, the voice may travel over the public network as switched calls and the data may travel over a private data line to another of the end-user's locations.

Fixed Capacity

Dedicated services are leased or built with a fixed capacity or bandwidth. For example, circuits are ordered from carriers such as AT&T or LDDS WorldCom at specific speeds and numbers of paths, or channels. Most organizations now lease lines with a minimum of 56,000 bits per second because of the minimal price difference between 56 kilobits and slower lines.

Examples of speeds of dedicated lines are:

- T-1: 24 channels, 1.54 megabits per second.
- Fractional T-1: 2 to 12 paths in increments of 56 or 64 kilobits per path.
- T-3: 672 channels, 44 megabits per second.
- Fractional T-3: 2 to 18 T-1s in increments of 1.54 megabits.
- 9600 bits per second.
- 19,200 bits per second.
- 56,000 bits per second.
- 64,000 bits per second.

Speeds of greater than T-1 are available for large organizations with high bandwidth requirements. T-3 links provide 672 simultaneous voice and/or data paths at 45 million bits per second. (T-3 is discussed in Chapter 6.)

An example of a dedicated T-3 link is the line between an exchange carrier and a customer such as an insurance company. Organizations such as insurance companies rent high-capacity T-3 links for customer service call centers that receive a high volume of calls. The dedicated portion of the link is between the customer and the carrier. Calls arrive at the carrier central office and are passed on a dedicated T-3 link to the insurance company. Service providers such as Internet service providers lease or own T-3 circuits in their own networks to carry Internet traffic.

Analog or Digital

Carriers and local phone companies sell both analog and digital private lines. Most end-users specify digital private lines because of superior quality, higher speeds and reliability compared to analog private lines. Digital lines also have more capacity and fewer errors. Speeds above 19,200 bits per second are available only in digital formats. Carriers and local telcos prefer to lease customers digital rather than analog lines as they fail less frequently than analog lines and are thus cheaper for vendors to maintain for customers.

Why Private Lines?

Security

Some organizations use private lines to transmit proprietary information or financial data. They feel private lines cannot be as easily tapped into and listened in on as switched lines. Organizations concerned about security may place encryption devices on both ends of dedicated services. The encryption device scrambles the transmission when it leaves the sending location and unscrambles it when it arrives at the receiving location.

Cost Savings

This is the major reason why firms lease private lines. If an end-user places more than two to four hours a day of voice or data traffic between two points, it may be cheaper to install a private line at a fixed cost rather than pay hourly costs. Consider the following hypothetical example of a switched call at $6 per hour:

$6/hour × 4 hours/day × 21 days/month = $504 per month
Monthly cost of a dedicated line = $400

At these prices, the organization may opt for the lower cost dedicated line. Many large firms have complex arrangements of dedicated lines called private networks. These private lines may connect many of the corporate locations for transmission of both voice and data communications.

Convenience and Functionality

For voice calls, end-users may want the ease of abbreviated dialing and possibly transparent features between sites. For example, telephone systems connected by private lines may have software installed allowing users to call each other using four or five digits instead of the eleven digits of a standard telephone call. The on-site telephone systems may also allow one set of operators to answer calls for multiple locations, which is a significant cost savings versus full-time operators at each location. Some telephone systems offer telephones with displays telling a user the name of the person from another corporate private network location that is calling him or her.

Transfer Between Sites Capability

One factor in the decision to lease private networks is the transfer capability between telephone systems connected by private lines. The ability to transfer calls to company locations over private lines enhances customer service. Customers without private lines connecting their sites cannot readily transfer callers between sites.

To illustrate, one newspaper chain grew by buying up small newspapers. They consolidated circulation and classified sales into two locations. However, customers continued to call incorrect offices to reach these functions. The organization installed private lines to connect all of its offices so that operators could seamlessly transfer callers to circulation and classified sales from all locations. This improved customer service because customers were no longer told to hang up and redial if they reached the wrong location.

The following are examples of applications for dedicated services:

• Video transmission to multiple sites within the same organization.

• The transfer of customer calls among multiple sites within a company.

• The ability for employees at multiple sites to access computer data and easily communicate via voice, video and email regarding joint projects.

• The transmission of customer orders between sales and factory or warehouse locations.

• Bulk transmission of images such as x-rays and customer applications for loans.

- Joint product development by employees in different time zones to turn out products faster. For example, a group in the Far East works on the same project at different times than a group in the U.S. The goal is to bring new products to market faster.

- Database access between distant sites for order status and technical updates on products.

Topologies of Dedicated Lines—The View from the Top

The term topology refers to the geometric shape of the physical connection of the lines in a network, the view from the top. It is the shape of the network a person looking down from above would see. The shape of the network, the configuration in which lines are connected to each other, impacts cost, reliability and accessibility. Consider the multipoint configuration. The multipoint design is used in polling environments where a computer such as a mainframe, polls terminals, one after another, asking if they wish to transmit or receive data. The computer polls the terminals in a predefined order. If one of the links in the multipoint network goes down, the entire network is out of service.

The following are private line configurations:

- *Point-to-point:* two tin cans and a string connecting two locations.

- *Multipoint:* more than two tin cans and a string connecting more than two sites together. In local area networks, this is also known as a bus design.

- *Star configuration:* all locations connect to, "hub into," a central site.

- *Mesh design:* all points on the network, nodes, connect to each other in a flat or non-hierarchical manner.

Figures 5.3 through 5.6 illustrate the number of lines used in sample configurations of each topology. Higher numbers of lines result in higher monthly charges.

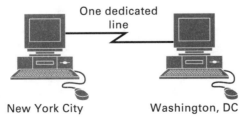

Figure 5.3 Point-to-point private lines; one private line connecting two locations.

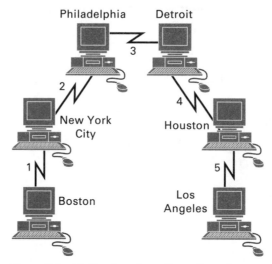

Figure 5.4   Multipoint private lines; five private lines connecting six locations.

The star shape is a common private network topology. In the case of a star configuration, all locations are connected to one main location (hub). If the main location goes down, all nodes (locations) on the network are disconnected from each other.

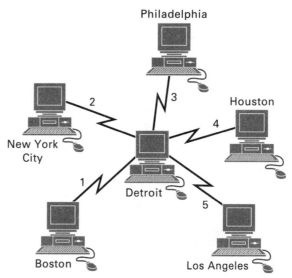

Figure 5.5   Star configuration; five lines connecting six locations. Each point must travel through the central hub to reach another point.

A mesh design results in higher costs, but has more reliability built in than a star arrangement. A mesh configuration results in more paths for information flow. If one link goes down, voice and data can be transmitted via other paths. The customer pays for more links in a mesh configuration, but has higher reliability.

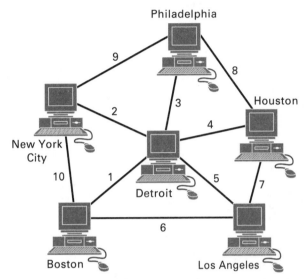

Figure 5.6   Mesh configuration; ten lines connecting six locations via a mesh configuration provides built-in redundancy but adds costs for additional lines.

Pricing

Rates are based on *distance and speed*. The higher the speed and the further apart the ends of the line, the higher the cost of the circuit. For example, a line at 19,200 bits per second between New York City and Boston may cost $400. The same speed line from California to Boston may cost $1800. Lines capable of carrying bits of data at faster speeds cost more than slow lines. A T-1 line at 1.544 million bits per second costs more than a line that can carry 56,000 bits per second. Volume and term discounts are available.

Pricing for interstate, inter-LATA, dedicated lines consists of two items (see also Figure 5.7):

- *Local channels:* Local channels run from a customer premise to the carrier's network. One channel is required at each end of the private line. For the most part, local channels are provided by local telephone companies and CAPs. Interexchange carriers are starting to provide local channels for the dedicated lines they sell.
- *Interexchange miles:* Interexchange mileage is within a carrier's network. The mileage runs from the access point, where it enters the carrier's network, to the egress point, where it leaves the carrier's network.

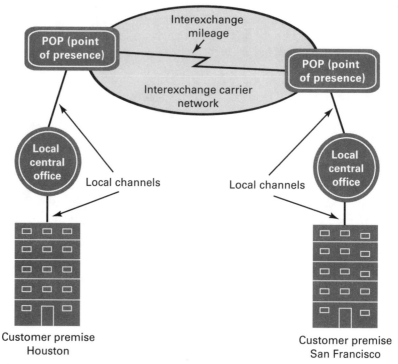

Figure 5.7   Local and interexchange legs of a private line.

In considering pricing on private networks, a major factor is the cost of organizational staff to maintain the lines and equipment used to connect individual users and computers to the wide area network or private lines. Organizational expertise is required from the initial design and sizing of the dedicated lines to the ongoing maintenance and addition of new applications requiring new demands on the dedicated network. For these reasons, outsourcing is popular as a way to manage the implementation, growth and data maintenance of private networks made up of dedicated lines.

Outsourcing to Manage Private Networks

Outsourcing is a "hot" industry. Companies often hire outside expertise to manage both their network services and equipment. Expertise is required for the selection of multiplexers, routers and modems that connect computers to networks. Ongoing tasks range from maintaining software for adding and deleting PC addresses in routers to determining the cause of failures on private lines. Companies expend energy in sorting out claims by network providers that equipment is faulty on customer premises and claims by equipment vendors that the outside lines are faulty. The goal of contracting with an outsourcing service is to have one vendor responsible for problem determination regardless of the location of the breakdown. Growth in outsourcing network and on site networking equipment maintenance is the result of the complexity required to maintain large corporate private networks. Use of virtual network services is another outcome of the complexity of maintaining, sizing and installing private networks.

Virtual Networks—Features of Private Networks without the Hassle

The demand for network services is growing exponentially, fueled by applications such as Internet access, file sharing, email and video teleconferencing. When organizations need to connect more than three or four sites to each other, repair and capacity issues on theses lines become burdensome. One solution is the implementation of virtual networks. A virtual network acts like a private network, as shown in Figure 5.8. It has many of the features of a private network. However, it is managed by a carrier. The customer connects dedicated or dial-up lines to the carrier's network. The carrier is responsible for connections between the customer sites. A dedicated link is not needed to each site. Links from each site to the carrier network do the job.

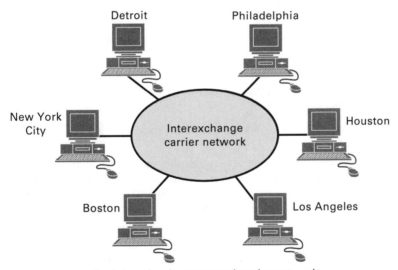

Each location is connected to the network
rather than directly to another location.

Figure 5.8   A virtual private network.

One example of a virtual network is a frame relay service. Frame relay technology is discussed in Chapter 9. Because of the high cost of private networks in both staff and lines, many organizations have a mix of private lines for routes with the highest amount of voice and data traffic, and switched services and virtual private network services for routes with less voice and data traffic.

Network Intelligence and Signaling

Overview of Signaling

People who call other people generally use the public switched telephone network. The caller dials a number and hears progress tones such as dial tone, ringing, busy signals or reorder tones. These are all signaling tones carried within the public network. In addition to tones, callers may hear digital messages telling them the number they dialed is not in service or has been changed.

The signaling innovations covered in this section represent a major improvement in the public network's capabilities. They not only enable carriers to manage their networks more efficiently, but they also provide the means for the introduction of profitable, new services.

Signaling innovations led to:

- Cost reductions and cost stabilization in the price of long distance.
- A platform for new, network-based services such as fax-on-demand, voice mail and interactive voice response.
- 800 and 888 number portability between carriers (the ability for an organization to retain its toll-free numbers when changing carriers).
- Lowered barriers to entry into the carrier market.
- Improved network reliability.
- ISDN.
- New "smart" services such as caller ID, call trace, call return and call waiting.

In addition to the above innovations, signaling is the backbone for interconnection between cellular, global wireless and multiple carriers' networks. The architecture of the signaling network was established by AT&T in conjunction with Bell Labs in the 1970s, before the 1984 divestiture. Prior to the 1984 divestiture, AT&T owned all of the 22 Bell Operating Companies. It had the necessary control of the public network that enabled it to set a standard that was followed across the country and later adopted by the international community.

The major innovation was that signals used to set up and tear down calls were to be carried on a separate path from voice and data conversation. (Examples of signals include: ringing phones, busy signals and the signals resulting from dialing a call.) This architecture of carrying signals on a separate channel from calls is called common channel interoffice signaling.

Background

Common channel interoffice signaling, also known as out-of-band signaling, is in reality a data communications network laid over carriers' switching networks that has opened markets for new products and enhancements to carriers' features. Common chan-

A Signaling Tutorial

Signaling is the process of sending information between two parts of a network to control, route and maintain a telephone call. For example, lifting the handset of a telephone from the receiver sends a signal to the central office saying, "I want to make a phone call." The central office sends a signal back to the user in the form of a dial tone, which says the network is ready to carry the call.

The three types of signals are:

- *Supervisory signals:* Supervisory signals monitor the busy or idle condition of a telephone. They are also used to request service. They tell the central office when the telephone handset is lifted, off-hook requesting service or hung-up, on-hook and in an idle condition.

- *Alerting signals:* These are bell signals, tones or strobe lights which alert end-users that a call has arrived.

- *Addressing signals:* These are dial tones or data pulses which tell the network where to send the call. A computer or person dialing a call sends addressing signals over the network.

Signals can be sent over the same channel as voice or data conversation, or over a separate channel. Prior to 1976, all signals were sent over the same path as voice and data conversation. This was called in-band signaling. In-band signaling resulted in inefficient use of telephone lines. When a call was dialed, the network checked for an available path and tied up an entire path through the network before it sent the call through to the distant end. For example, a call from Miami to Los Angeles tied up telephone capacity in multiple locations and a path throughout the network after the digits were dialed, but before the call started.

Estimates vary that from 20% to 35% of calls are incomplete due to busy signals, network congestion and ring-no-answers. Therefore, channels that could be used for voice or data conversations were used to carry in-band signals such as those for incomplete calls, dial tone and ringing. Multiplying this scenario by the millions of calls placed resulted in wasted telephone network facilities.

In addition to tying up telephone facilities, in-band signaling set up calls more slowly. To illustrate, with in-band signaling, the time between dialing an 800 call and hearing ringback tones from the distant end may have seemed lengthy to the caller. This is the call setup part of the call. *Call setup* includes dialing and waiting until the call is actually established. Call setup is slow with in-band signaling. Carriers do not bill for call setup and thus use valuable network capacity for call setup without getting revenue. Carriers do not receive revenue until the actual connection of a telephone call.

nel signaling started as a way to increase the efficiency of a network by setting up separate channels for signals. It evolved into the basis for intelligent networks. Routing instructions, database information and specialized programs are stored in computers in the carriers' networks and are accessible over out-of-band signaling links.

Beginning in the late 1970s, the public network evolved from purely carrying voice and data calls to a vehicle with intelligence, greater capacity and faster recovery from equipment failure. The impetus to upgrade the network came from AT&T's desire to more cost-effectively manage and add capacity to the network. This upgrade laid the foundation for new services such as enhanced 800 services, ISDN, call forwarding, three-way calling and call waiting.

Increased efficiency, automation and functionality evolved out of a new way to carry signals such as ringing, dialing and disconnects, which was developed by AT&T in the 1970s. The new signaling, called common channel interoffice signaling, carries signals on a separate data communications network rather than on the same path as a person's voice. Out-of-band signals are faster than in-band signals and use less network time for non-billable call setup time.

Signaling System 7 (SS7)—Lowering Costs and Increasing the Reliability of Public Networks

The Signaling System 7 (SS7) protocol, which is based on common channel signaling, is a factor in lowering barriers to entry into the common carrier market. Routing intelligence is being migrated from more expensive central office switches to lower cost computer-based peripherals. For example, Stratus sells parallel processor computers to carriers capable of holding massive databases with information such as routing instructions for 800 calls. One processor with its database supports multiple central office switches. In this case, each central office switch is not required to maintain sophisticated routing information. The expense of the upgrade is shared among many central offices.

The significance of advancements in signaling technology should not be underestimated. If problems are detected, the information is sent over the signaling network to centralized network maintenance centers where technicians see visual indications of alarms on wall-mounted, computerized display boards. Moreover, sections of carrier networks can be quickly reconfigured from commands sent by centralized network control centers over signaling channels. During the earthquake in California, the network was reconfigured such that no calls were allowed into California. This left paths open for Californians to make calls out of the state to reassure relatives that they were safe.

Adding Features to Carriers' Networks

Carriers communicate information between specialized servers and central office switches over out-of-band signaling network links. For example, databases located in carriers' public networks are capable of storing the profiles of both the telephone number dialed and the telephone number of the caller for each call. This intelligence enables firms to customize call treatment. For example, from 8 AM to 5 PM Eastern Standard Time, 800 calls may be routed to the east coast. After 5 PM Eastern Standard Time, all calls may be routed to California.

Because the databases and specialized processors are on separate computers from the central office switch, the central office and its switch do not have to be upgraded each time new "intelligence" is added to the network. This is a significant factor in allowing faster additions of features to the public network. In addition, the specialized servers and databases are accessed by multiple central offices. Thus, adding upgrades through the signaling network provides additional functionality to multiple central offices.

Examples of features available on networks with common channel signaling include:

- Telephone and cellular company voice mail with signaling links between voice mail systems and the central offices.

- Voice-activated dialing for calling cards, car phones and home phone lines supported by speech recognition systems in the network.

- Automated "roaming" on cellular telephone networks, where roaming locations are kept on computers in the network.

- Multiple telephone numbers associated with one residential telephone line such that databases contain information on ringing cadences, e.g., one ring for line xxx-1234 and two rings for line xxx-3456 for each telephone number.

- Custom calling features from local telephone companies such as call forwarding, call conferencing and call waiting.

- Load balancing by call volume, e.g., 50% of the calls sent to California and 50% to the call center in Iowa.

- Calling number and calling name delivery (display of the calling telephone number and name associated with the telephone placing the call).

- Customer links to carrier networks, where customers specify new call destinations for the 800 services into their call centers. For example, calls may be redirected based on unusual traffic patterns.

- Personal numbers that "follow" a person from place to place, so that the person is reachable from the same phone number when they are in their car, at work and at home.

Residential, business and cellular voice mail purchased from local telephone companies is an example of a service made possible by common channel signaling. Single voice mail systems manufactured by vendors such as Octel and Boston Technology support multiple central offices. The following are some of the messages communicated between network-located voice mail and central office switches (see Figure 5.9):

- Initiate stutter dial tone or light message waiting indicators to tell users they have voice mail messages.

- Instruct the voice mail system on which person should receive specific voice messages, e.g., play Mr. Smith's greeting for this call.

- Turn stutter dial tone and message waiting indicators off when voice mail messages have been listened to.

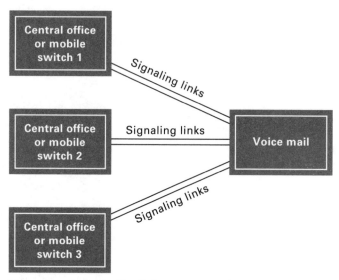

Figure 5.9   Central office links to voice mail.

Common Channel Signaling, Efficiency and Redundancy

Part of the efficiency of common channel signaling is due to one signaling link supporting multiple voice and data conversations or paths. This is a separate data network that carries all of the signaling in each carrier's network. The fact that one signaling link supports many trunks (transmission paths between telephone switches) highlights the requirement for reliability. If one signaling link crashes, many routes are out of service. Redundancy is an important consideration in the design of carrier networks (see Figure 5.10 for an illustration of redundant signaling links).

Local telcos have two types of central offices. Tandem offices work in tandem or along with each central office. They connect central office to central office and central office to interexchange carriers' switches. Tandem offices do not have connections directly to end-users. Signaling System 7, SS7, is used throughout the tandem office portion of local networks. These offices carry high volumes of calls on paths called trunks.

The other type of central office in local networks is an end office. End offices have many analog, "plain old telephone" lines between homes and telco switches. In these analog lines, the dialing sequence is sent over the same channel as the user data or call. The volume of calls from end offices to homes and from end offices to tandem offices is lower than on trunks between tandem offices. End offices are the last switches in the public network to be upgraded to SS7.

Consider calling line ID, also known as CLID. With calling line ID, the called party sees the telephone number, and with some equipment, also the name of the person calling. The calling line ID, or identity of the calling party's telephone number, is transmitted over

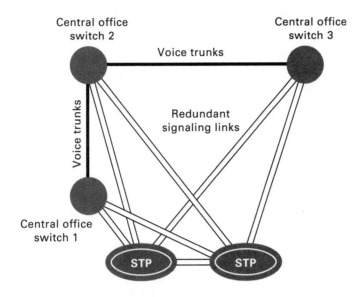

Central office
switch 2

Central office
switch 3

Voice trunks

Voice trunks

Redundant
signaling links

Central office
switch 1

STP   STP

STP : Signal transfer points
route signals over paths
separate from the voice calls

Figure 5.10 Common channel signaling.

a separate path from the call itself on the tandem central office to central office portion of the call. However, on the portion of the call from the end office to the customer, caller ID is sent on the same channel as the call.

ISDN is a digital telephone service for voice and data calls. With ISDN, user data and voice use a separate channel from signals in all parts of the network, including the end office to the customer. Signals such as the number dialed, dial tone and ringing are sent as bits over the separate signaling channel. This is why data users find calls are set up so quickly with ISDN. ISDN was made possible with SS7. It is faster than POTs, plain old analog, telephone service. ISDN service is discussed in Chapter 7.

Signaling System 7—The Glue for Links between Carriers

A major value of Signaling System 7 is its ability to enable all carriers to work in concert with each other. It is a standard protocol approved by the ITU, International Telecommunications Union. Signals are sent from central office to central office, from local exchange carriers to interexchange carriers, from domestic carriers to international phone companies and from central offices to specialized processors and databases. Global billing, 800 services and international roaming for wireless calls are all dependent on SS7.

Early SS7 Implementations

The Swedish PTT trialed SS7 in 1983. The United Kingdom and France also had early implementations in the early 1980s. MCI first implemented SS7 in their network in April of 1988 in Los Angeles and Philadelphia. According to a May 2, 1988 article in *Network World*, MCI stated that it had halved call setup time on calls over the Philadelphia/Los Angeles route. Freeing up voice paths from signaling enabled carriers to pack more calls on their existing network paths.

Signaling System 7, SS7, is used with variations throughout the world and is based on common channel signaling architecture. With CCS, common channel signaling, signals can be sent in both directions at once, resulting in faster speeds. Access is allowed to databases and specialized processors for applications such as fax mail, voice mail, order entry, local number portability and distribution of calls among multiple call centers.

The precursor to SS7, Signaling System 6, was developed by AT&T in the 1970s for the old Bell System. It was thought that in-band signaling tones were too limited, wasted network capacity and set up calls too slowly. AT&T felt that it could take advantage of computer-controlled switching to develop an overlay signaling network that could pass complex messages. In essence, it could create a data communications network that could pass more complex messages than the limited in-band tones which notified the network when calls were completed, how they were addressed, etc. Signaling System 6 links initially transmitted signals at 2,400 bps and later at 4,800 bps. The first implementation of Signaling System 6 in 1976 by AT&T was automation of calling card validation. This automation enabled AT&T to eliminate operators for calling card validation.

Prior to the addition of common channel interoffice signaling in the network, operators were required to set up all calling card calls. With common channel interoffice signaling, authorization for calling card calls is done automatically by checking the telephone company databases called LIDBs, or line information databases. Line information databases contain all valid telephone and calling card numbers.

With CCS, when a user makes a calling card call, an operator is not required to check a computer database to determine if the calling card number is valid. Instead, the central office sees from the number dialed that the call was made from a calling card. It then initiates a call to the LIDB to determine the validity of the calling card. An operator is not required to perform this validity check. The function is automated.

AT&T mandated the implementation of Signaling System 6 throughout the entire Bell System in the late 1970s. The faster SS7, a layered protocol with signaling links of 64,000 bps was specified by AT&T in 1980. It was approved by the CCITT (in French, International Telegraph and Telephone Consultative Committee) as an international standard in 1980. (The CCITT is now called the ITU, International Telecommunications Union.) It is used, with variations, on a global basis in every country.

Many people feel that a network structural change as major as that required to agree upon Common Channel Interoffice Signaling and Signaling System 7 would be much more difficult in today's environment. Agreement would have to be reached between major interexchange carriers, local telephone companies, cable companies and competitive access providers (CAPs). In the 1970s, AT&T owned all of the Bell Operating Companies and could set the direction, networkwide, for the whole company, at that time, AT&T plus the 22 Bell Operating Companies.

As with many standards, implementation of SS7 differed among countries. For example, the U.S., Canada, Japan and bits of China implemented the ANSI, American National Standards Institute version of SS7. Europe implemented the ETSI, European Telecommunications Standards Institute, version. ANSI is a U.S. voluntary standards setting board. ETSI is its European counterpart.

To enable wireless and standard carriers to communicate over SS7 with international carriers, companies such as TsDesign in Atlanta, Georgia produce software designed to operate on PCs that enable American-manufactured central office switches to communicate, for instance, with Chinese and European SS7 implementations.

SS7 Components

See Figure 5.11 for an overview of the following SS7 components.

Signal Transfer Points (STPs)

Signal transfer points route signals between central offices and specialized databases. Signal transfer points are packet switches. Messages are sent between points on the SS7 network in variable-length packets with addresses attached. (Think of the packets as envelopes of data with user information such as the called and calling telephone number, error correction information and sequencing numbers so that the correct packets or envelopes are grouped together in the correct order at the receiving end.) Signal transfer switches read only the address portion of the packets and forward the messages accordingly.

The fact that the signals are sent in a packet format is a significant factor in SS7's capability of one link being able to support multiple call paths. Packets from multiple "conversations" share the same pipe. Packets from conversations "a", "b", "c" etc. are broken into small chunks (packets) and sent down over the same 64,000 bps SS7 links with each other.

Service Switching Points (SSPs)

Service switching points enable central offices to initiate queries to databases and specialized computers. Service switching points are software capable of sending specialized messages to service control points (databases). For example, when a 900 call is dialed, SSPs set up a special query to a 900 database (the service control point) for information on routing the call.

Service switching points convert the central office query from the central office "machine language" to SS7 language. When signals are received from the signaling network, the service switching points convert the SS7 language to language readable by the central office switch.

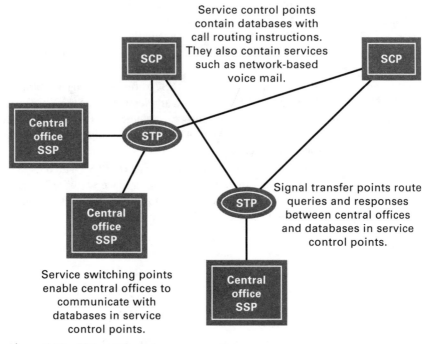

Figure 5.11 SS7 components.

Service Control Points (SCPs)

Service control points hold specialized databases with routing instructions for each call based on the calling party and/or the called party. For example, the service control point tells the network to which carrier to route an 800 call, e.g., AT&T, MCI or the local telephone company. In cellular networks, the SCPs maintain the roaming information. They communicate information to the wireless carrier as to within which region the cellular telephone is located.

New services such as network-based voice mail, fax applications and voice-activated dialing are located on service control points or intelligent peripherals.

Migration to Higher SS7 Speeds

In the future, to gain more capacity, the speed of SS7 links may be upgraded to T-1, 1.54 million bits per second, and after that, to ATM speeds, 155 million bits per second. An example of an application for high-speed SS7 includes the setup by cable companies for real-time customer requests for movies.

Advanced Technologies, the Internet and Wireless

CHAPTER 6

Specialized
Network Services

$\mathbf{T}$his chapter explores high-speed, digital network services available to telephone companies, long distance companies, Internet service providers, corporations, telecommuters, non-profit organizations and high-end residential users of services such as the Internet. The intent is to explain how the services work, explore the applications and discuss where the services fit. A summary of the services is listed in Table 6.1.

A characteristic that all of the technologies in this chapter have in common is that they are all digital transmissions. As discussed previously in this book, analog services inhibit the higher speeds, accuracy and reliability provided with digital communications. Telecommunications is booming. End-users want more telephone lines and higher speeds for applications such as facsimile, access to the Internet and work-at-home applications. Some of the services reviewed in this chapter have been available for some time. These include T-1 and T-3. T-1 allows 24 voice or data calls on two pairs of copper, fiber or microwave. T-3 has the capacity for 672 channels and a speed of 44 megabits. ISDN, integrated services digital network, has had limited availability for a long time and is just becoming more readily available from carriers. Basic rate ISDN gives users two telephone lines on one pair of wires. Each of the lines can be used for voice or data. If used for data, the speed and call setup are faster than analog lines used with modems.

One technology that has grown considerably in the past five years is frame relay. Frame relay is a value-added data network used mainly to connect distant local area networks together. Each of an organization's locations that wishes to use the frame relay service has a line to the carrier's frame relay network. Locations do not have to maintain a line and data communications devices for connections to every other site. The carrier takes care of planning for capacity and network reliability.

In addition to T-1, T-3, frame relay and ISDN, there are a myriad of new services available or discussed in the media and introduced by vendors at sales presentations.

Table 6.1   An Overview of Specialized Digital Network Services

Network Service	Places Typically Used	How Used
T-1 24 voice or data channels	Medium to large organizations	Access to Internet and long distance companies' backbones
T-3 672 voice or data channels	Large organizations, Internet service providers, telcos	Backbones and access to long distance companies
BRI ISDN Two voice or data channels + one signaling channel	High-end residential customers, organizations	Video conferencing, telecommuting, Internet access
PRI ISDN 23 voice/data + one signaling channel	Internet service providers, PBXs, automatic call distributors (ACDs)	Call centers, room-sized video conferencing, remote access to corporate databases
Digital Subscriber Lines, DSL 128 Kbps to 6 megabits	Telecommuters, corporations, Internet service providers, high-end residential customers	Remote access to corporate databases and Internet access; some types used for voice
Frame Relay 56 Kbps to 45 megabits access to value-added data networks	Medium to large commercial customers	A public, primarily data network service for local area network to local area network connections
ATM 56 Kbps to 622 megabits	Telcos, Internet service providers, frame relay networks, and large organizations such as major universities	Used to switch high-usage backbone voice, video and data traffic
SONET Up to 129,000 channels on fiber optic cable	Telcos	Transports customer traffic at standard speeds over fiber; available in the local loop to provide extra reliability to large corporate customers

These are high-speed services and their acronyms sound like alphabet soup: ATM, SONET and DSL. To non-technical professionals, there is an aura of mystery surrounding these technologies. The goal of this chapter is to sort out the meanings, capabilities and applications of these services.

ATM, asynchronous transfer mode, switches carry voice, data, image and video at very high speeds of up to 622 megabits over fiber optic cabling. ATM is used mainly in Internet networks, carrier networks and frame relay networks. It is also starting to be used by very large universities, financial organizations and Fortune 100 companies to meet their needs for high-capacity telecommunications services.

SONET, synchronous optical network, is also a high-speed telecommunications service that works on fiber optic cabling. SONET runs at speeds of up to ten Gigabits. It is used in local and long distance company networks to carry traffic from multiple customers running at different speeds. For example, SONET networks can accept T-3 traffic from customer A and T-1 traffic from customer B and transport both streams at SONET speeds to their destinations. SONET provides a way for carriers to increase the capacity of their networks.

Finally, there is a "family" of DSL, digital subscriber line, services. DSL technology is a way to increase the speed of the copper cabling in the local loop without adding new copper cabling. The local loop is the portion of the public network between the telephone company switches and end-users' locations. This often analog "last mile" of the public network is the main bottleneck to providing consumers and businesses with high-speed telecommunications lines for options such as Internet access and work-at-home applications. There are a variety of DSL services. They run at speeds of from 128 kilobits to 52 megabits. The most common speed is 1.54 megabits.

T-1—24 Voice or Data Paths over One Telephone Circuit

T-1 services were developed by AT&T in the 1960s to save money on the telephone companies' outside cabling. Instead of continuing to lay expensive copper cabling, a scheme was devised to carry 24 voice or data conversations over one telephone circuit. (A circuit is a path for electrical transmissions between two points.) T-1 was envisioned as a way to save money on cabling for calls carried between telco switches. The technology was not made available directly to end-user locations until 1983.

In the mid to late 1980s, large organizations such as universities, financial institutions and Fortune 100 companies used T-1 circuits to tie locations to host computers for applications such as order entry, payroll and inventory. The T-1 often replaced having to physically carry large computer tapes between locations. Once installed, companies found the T-1 to be light-years ahead of old analog data lines in terms of reliability. The digital form in which T-1 is installed greatly enhances its reliability compared to older analog telephone lines.

When T-1 first became available, only the very largest organizations could justify paying its, at that time, high rates. Not only were rates high, but the service itself took months to install. Until competitors such as alternate access providers and interexchange carriers such as MCI and Sprint offered T-1, the lead times often stretched out from six months to a year. A major problem in provisioning T-1 was the fact that the service, which is digital, often needed to be connected to older, analog telephone switches. This was accomplished through using channel banks between the T-1 line and the central office switch. Channel bank charges are passed on to customers, making their use an extra cost. The channel banks are also another point of possible failure.

Channel Banks—Connecting T-1 to Analog PBXS and Central Offices

Channel banks, as illustrated in Figure 6.1, are the multiplexing gear that connect digital T-1 circuits to analog PBXs and central office switches. The channel bank takes the signals from analog systems such as older analog PBXs and samples each of a possible 24 analog voice or data streams 8,000 times each second. It digitizes these voice and data connections and sends them down the digital channel. At the receiving end of the T-1 circuit, the channel bank decodes digital signals back to analog. The methodology used for sampling and coding these signals is pulse code modulation, PCM. Decoders within the channel bank perform coding and decoding functions of converting analog voice to digital, and vice versa.

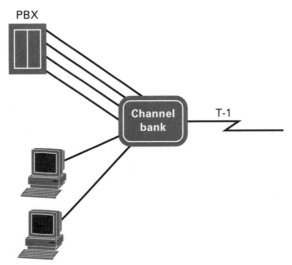

Figure 6.1   Using a channel bank so that a T-1 circuit can be shared for voice and data.

Like most other equipment, channel banks have shrunk considerably in size. Early channel banks were the size of a china cabinet; newer banks are comparable to the size of an encyclopedia and can be hung on a wall.

While most local telephone switches are digital, channel banks are still required when customers order T-1 circuits served by older analog central offices. In some cases, customers with digital PBXs can connect T-1s directly into their digital PBX, but the central office portion of the T-1 requires a channel bank. Telcos charge customers a surcharge when a channel bank is needed because the central office is analog.

Channel banks are used today with digital as well as analog PBXs. A channel bank is one way customers share the T-1 for voice as well as data. The channel bank takes a number of the 24 channels and sends them to computing devices and the rest as analog transmissions to analog trunk equipment within a PBX. (Digital PBXs terminate analog as well as digital telephone trunks. Telco lines are called trunks when they terminate on a PBX.)

In addition to breaking out T-1 channels between data devices and voice telephone systems such as key systems and PBXs, channel banks can break out individual T-1 channels into lower than 64 kilobit speeds. These lower than 64 Kbps speeds are known as sub-rate speeds. Sub-rate channel speeds are 2.4, 4.8, 9.6 and 19.2 Kbps. The total on the sub-rate channels for a single 64 Kbps channel cannot equal more than 64 Kbps.

DS-0 and DS-1—64,000 or 56,000 vs. 1,544,000 bps

- DS-0 = 64,000 bps
- DS-1 = 1,544,000 bps

The letters "DS" above stand for digital signal level. This is the speed at which the various T-1, T-3 and fractions of these speeds run. People often refer to T-1 lines as DS-0 and DS-1. The DS-1 is the entire bandwidth of the circuit, 1.544 megabits. DS-0 refers to the speed of each channel of the T-1 circuit, 64 Kbps. The entire T-1 can be purchased, or single digital lines at 64 Kbps can be purchased. A full T-1 becomes lower in cost when organizations have from six to eight DS-0 lines.

Note that although each channel of the T-1 is 64,000 bps, the entire bandwidth of the circuit is higher than 24 x 64,000, which equals 1,536,000. The extra 8000 bits (1,544,000 - 1,536,000 = 8,000) are used for synchronization, keeping the timing set between frames. A frame is a transmission where bits from each of the 24 channels have been sampled and put onto the T-1 line.

Often, organizations want digital connections between their locations, but they don't have enough traffic to warrant paying for a full T-1 line. For example, a T-1 line between New York City and Boston costs about $5,000 per month. A digital line that runs at the 56,000 bps speed may only cost about $800 monthly. A publishing organization in Boston with a sales/marketing office in New York has a 56,000 bps line connecting the two offices. The line is used primarily for email and transmission of sales proposals between the two sites and the 56 Kbps capacity is sufficient at present.

All DS-0s run at 64,000 bps. However, depending on the signaling available in the telephone company's network, 8,000 of the bits may be required for signaling and maintenance functions, leaving only 56 Kbps for user data. Clear channel signaling must be available with the chosen carrier to be able to use the full 64 Kbps for user data. With clear channel signaling, the 8,000 bits don't have to be "robbed" for network maintenance. Thus, the full 64,000 bits are available for user data.

Media over which T-1 Signals Are Transmitted

T-1 can be installed on a variety of media. T-1 works with all of the following:

- Fiber optics.
- Twisted pair.
- Coaxial cabling.
- Microwave.
- Infrared light.

In the 1980s when T-1 was first installed for large corporations, the media it was installed on was mainly twisted pair copper. When T-1 circuits are run over twisted pair, two pairs (four wires) are used. One pair is used for transmitting and one pair for receiving. Telephone companies who used T-1 in its early days, the 1960s and 1970s, used a combination of coaxial cables, twisted pair and microwave. Microwave worked well for hard-to-cable areas such as the Grand Canyon.

As fiber optic cabling became available in the 1980s, its lightweight, high-capacity and low maintenance requirements made it a good choice for telcos. As more and more fiber optics are being used in local telco networks, the local phone companies are often bringing fiber directly into users' premises when they order T-1 circuits. If fiber is brought into a user's premise, the end-user must supply the electricity for the equipment that converts the signals between electrical pulses for internal copper cabling and light pulses for the outside telephone company fiber. If there is no backup power, customers lose their T-1s if they lose power in their buildings.

European vs. American and Japanese T-1—24 vs. 30 Channels

The only digital signal speed that is standard throughout the world is the DS-0 speed of 64 kilobits. Speeds of DS-1 and above are different. For example, the U.S., Canada and Japan use 1.544 for T-1 with 24 channels, while the rest of the world uses 2.048 with 30 channels of capacity (see Table 6.2). People who want to run a T-1 from the U.S. to an office in Europe need rate adaptation equipment so that the carrier in the U.S. can connect the domestic T-1 to the European T-1 line. T-3 speeds are also different in North America, Japan and Europe. DS-0, DS-1 and DS-3 are the most prevalently used of the T carrier speeds. Some people refer to European speeds as E1, E3, etc.

Speeds higher than DS-3 are usually carried by telephone carriers at SONET and ATM rates (see end of this chapter).

An Explanation of Time Division Multiplexing and Its Limitations

All T carrier signals, e.g., T-1 and T-3, are based on time division multiplexing. Each of the devices which communicates over a T-1 line is assigned a time slot. If there are

Table 6.2   Digital Signal Levels

Level	N. America		Japan		Europe	
	Circuits	Speed	Circuits	Speed	Circuits	Speed
DS-0	1	64 K	1	64 K	1	64 K
T-1 (DS-1)	24	1.544 M	24	1.544 M	30	2.048 M
T-2 (DS-2)	96	6.312 M	96	6.312 M	120	34.368 M
T-3 (DS-3)	672	44.7 M	480	32.06 M	480	34.368 M
T-4 (DS-4)	4032	274.17 M	5760	400.4 M	1920	139.3 M

eight telephones contending for a T-1 circuit, a time slot is saved for each telephone for the duration of the particular telephone call. For example, telephone 1 might be assigned slot A, telephone 2 slot B, etc. Similarly, if PCs were contending for the T-1, PC 1 would be assigned time slot A, PC 2 would be assigned time slot B, etc. If a PC paused and did not send for a few minutes, the slot would not be assigned to another computer. The assigned time slot would be transmitted without any bits. This is why time division multiplexing is not an efficient way to use a wide area network. Pauses in data transmission result in idle time slots. In a network with millions of time slots, this can result in many idle time slots and wasted bandwidth (see Figure 6.2).

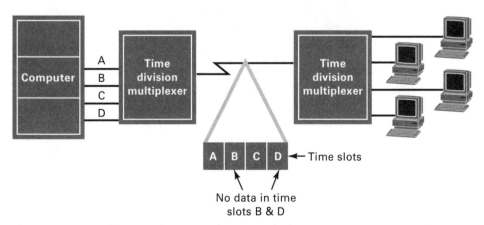

Figure 6.2   Time slots wasted in a time division multiplexing circuit.

Newer transmission techniques such as ATM do not assign specific time slots to each device. Rather, only transmitted bits use bandwidth. This results in a more efficient use of transmission capacity (see the end of this chapter).

A Sampling of T-1 Configurations Using T-1 for Combining Voice, Fax, Video and Data

To save costs on wide area networks, organizations combine voice, fax, video and data over T-1 circuits. The voice lines are usually connected to a PBX. Video conferencing unit connections vary; sometimes they go through a PBX, and sometimes they bypass the PBX. If they do transmit via a PBX, the PBX needs to be able to send data at high speeds. More often, however, organizations bypass the PBX for transmitting data. An option for these connections is to terminate the T-1 in a drop and insert multiplexer. The drop and insert multiplexer drops off some channels of the T-1 to data devices. It then "stuffs" bits into the channels it dropped off to the data devices and sends a full T-1 including the "bit stuffed" channels to the PBX's T-1 multiplexing equipment. See Figure 6.3 for a visual depiction of the drop and insert concept.

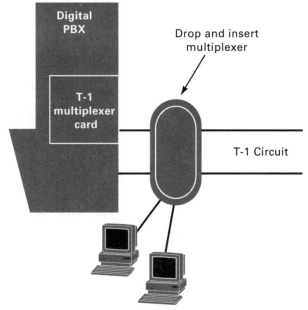

Figure 6.3 A drop and insert multiplexer with a digital PBX, so voice, data and video can share a T-1 circuit.

Digital Cross Connects—Flexible Capacity

Digital cross connect services are commonly used in organizations which have multiple sites they want to connect with private, dedicated lines. (Dedicated lines are for the exclusive, full-time use of a specific organization. They are run between specific, defined points. (See Chapter 5.) If an organization needs a full T-1 at headquarters but partial T-1s at its remote sites, one option is to use digital cross connections provided by the telco. With digital cross connections, the telephone carrier runs a full T-1 from its telephone switch to the organization's headquarters. The carrier then runs less than 24 channels to the other locations, as shown in Figure 6.4.

Some carriers will give customers the capability to reconfigure their T-1 channels from a terminal or PC at the customer's location. This is done with applications such as rerouting traffic in the event of a disaster or time-of-day peaks in data communications traffic.

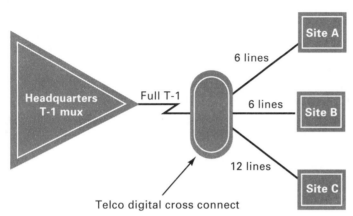

Figure 6.4 Digital cross connect switching used in a four-site wide area network. The digital cross connect enables the headquarters to change the configuration to, for example, eight lines to site A and ten lines to site C if traffic patterns change.

Fractional T-1—When 24 Paths Are Not Required

Customers who require more than 56 Kbps, but less than a full T-1, often opt for fractional T-1, less than a full 1.544 megabits. For example, a four-channel 4 x 64 = 256 Kbps or 4 x 56 = 224 Kbps costs less than 1.544 megabits. A customer using a 56 Kbps line between New York City and Boston for email and Internet access may at times need more capacity for video conferencing. Instead of ordering a full T-1 at $5,000 per month, the company should consider a fractional line with four channels for $2,000 per month. Fractional T-1 services are available in increments of two channels starting at 2 x 64 = 128 Kbps. Most organizations don't use more than about six channels for fractional T-1, as the price starts to get close to a full T-1 above six channels of a fractional T-1 service.

T-3—The Capacity of 28 T-1 Lines, 672 Channels

A T-3 circuit is equivalent to 28 T-1s, or 672 channels (28 x 24 = 672). The total speed of a T-3 line is 44.736 megabits. This speed is higher than 28 x 1.544 because some bits are needed for overhead, i.e., signaling and maintenance.

With the growth in demand for local and long distance services, organizations and telephone carriers are finding that they require several T-1 circuits to a single location. Call centers, reviewed in Chapter 2, are a prime example of business services with large calling volumes. Catalog sales, financial institutions, insurance companies and service bureaus that provide call center functions for smaller companies have large consolidated groups of agents that make and receive calls.

In addition to making and receiving calls from customers, many Fortune 100 companies also install telephone lines to carry voice, video and data traffic between their large sites. For both call centers and these high-traffic, backbone networks, many T-1 circuits are needed to keep up with heavy volumes. Rather than install multiple T-1s, companies with large calling and data communications volumes install T-3 circuits. T-3 services start to cost less than multiple T-1s when customers have between eight to ten T-1s at the same site.

Internet service providers, telephone companies and frame relay backbone networks are other groups that utilize the 672-channel capacity of T-3. Many telephone carriers, frame relay providers and Internet service providers are upgrading from T-3 to higher bandwidth ATM speeds. However, they often keep T-3 service as a point where customers access their network. For example, multiple customers call into the T-3 network entrance points from individual 56 Kbps to 1.54 T-1 lines.

ISDN—Integrated Services Digital Network

ISDN is a digital, worldwide public standard for sending voice, video, data or packets over the public switched telephone network. It enables customers to have one or two pairs of wires that they can use for voice, data and video brought onto their premises. There are two "flavors" of ISDN, basic rate, BRI, and primary rate, PRI, which are defined in Table 6.3.

Table 6.3   Basic Rate and Primary Rate ISDN

ISDN Service	# and Speed of Channels	Total Speed	# of Pairs of Wires from Telco to Customer Premise
BRI ISDN	3 total: 2 at 64 kilobits; 1 at 16 kilobits for signaling or packetized data	144 Kbps	1
PRI ISDN	24 total: 23 at 64 kilobits; 1 at 64 kilobits for signaling or packetized data	1.54 megabits	2

Both types of ISDN have these important characteristics:

- *Digital:* digital connectivity to achieve consistent, high-quality calling.
- *Out of channel signaling:* calls are set up quickly; each voice/data channel uses all of the bandwidth for user data. The signaling channel is available for packet switching.
- *A switched service:* Users pay long distance or local calling fees for the amount of time they use the line.
- *A standard interface:* All users with ISDN can communicate with each other.

ISDN standards were first published in the mid-1980s, but were finalized in the 1990s. A public network call over ISDN was first demonstrated by Nortel in 1987. ISDN works over existing copper wiring. It does not require fiber as do higher speed ATM and SONET services. However, central offices need ISDN-capable equipment. Some people feel that ISDN is an interim speed until XDSL, cable modems and other higher speed, megabit services are widely deployed. The total speed of BRI ISDN is 128 Kbps. PRI ISDN's top speed is 1.536 megabits.

An issue in acquiring ISDN has been its availability. ISDN has been gaining in popularity and availability in recent years. Telco employees attending Northeastern University's State-of-the-Art Program report that customer requests for BRI ISDN are increasing rapidly. Moreover, telco knowledge ability in installing ISDN is growing. However, overall deployment is still low. Experts estimate 1977 penetration at 5% of total telephone lines. Deployment of BRI ISDN is higher in Europe and Japan than the U.S. France, Germany, Japan and Switzerland are widely acknowledged to have a large base of BRI ISDN customers. The important point about ISDN is that availability is growing and increasing numbers of people are installing the service. A user must be within 18,000 feet, 3.4 miles, from the central office. According to a Bell employee source, 85% to 95% of telephone lines fit this criteria.

Basic Rate Interface ISDN—Two Channels at 64 Thousand bps

Basic rate interface, BRI, consists of two bearer channels for customer voice or data at 64 Kbps. In addition, it has one 16-Kbps signaling channel. It runs over a single pair of twisted wires between the customer and telco. Common ways people use BRI ISDN are:

- Internet access.
- Desktop video conferencing.
- Use of the D, signaling channel, for credit card verification.
- Downloading software from the Internet.
- Work-at-home applications.
- Connecting remote local area networks together.

Figure 6.5 illustrates the characteristics of a BRI ISDN line.

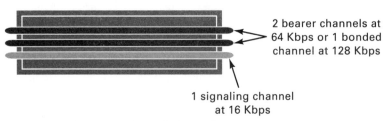

2 bearer channels at 64 Kbps or 1 bonded channel at 128 Kbps

1 signaling channel at 16 Kbps

Figure 6.5 Channels and speed of a BRI line.

Downloading Software from the Internet via BRI ISDN

Many software companies are distributing software upgrades, new releases of software and even the sale of new software by making them available for downloading from the Internet. They are saving money on discs, paper and postage. The amount of time it takes users to download this software depends on the speed of the modem they use. Table 6.4 illustrates the optimal amount of time required to download files via modems versus ISDN service. It is important to recognize that modem times depend on the quality of the telephone line. Often, a 28.8 Kbps modem does not achieve its optimal speed due to noise and interference on the line. Table 6.4 compares the speed of downloading a ten-megabit file. In the last line of the table, two ISDN bearer channels are "bonded" together, combined, to achieve a total of 2 x 64 = 128 thousand bps.

Table 6.4 The Comparative Time to Download a 10-Megabit File

Speed/Type of Service	Time to Download the File
14.4 Kbps modem	93 minutes
28.8 Kbps modem	46 minutes
1 ISDN bearer channel at 64 Kbps	21 minutes
2 ISDN bearer channels at 128 Kbps	10 minutes

Table information provided by NYNEX.

ISDN for Work at Home

As more users telecommute and open home-based businesses, they are finding that they need multiple telephone lines for voice, facsimile and Internet access. Corporate staff who set up telephone lines for telecommuters routinely order three telephone lines for fax, email and voice. People who have their own businesses often provide their own callers with poor telephone service in the form of busy signals and call waiting interruptions because they have too few lines. They try to juggle family telephone calling, email, fax and business calls on too few lines. They also waste their own time using slow modems to download Internet files.

There is growing interest in using ISDN to share one line for voice, data and facsimile. People who use an ISDN line need a terminal adapter with an NT-1, network terminating device. A terminal adapter with a built-in network terminating device is used to connect non-ISDN equipment to an ISDN line. (See Chapter 7 for information on NT-1 devices.) The terminal adapter converts the ISDN line from two wires coming into the customer's premise to four-wire inside wiring. Each non-ISDN device plugs into the back of the terminal adapter.

Video Conferencing Using ISDN

The price of desktop video conferencing has dropped markedly in the last two years. This has spurred sales of these units. The price of a desktop video conferencing system is $3500 and lower. Many organizations purchase these systems as a way to try out video conferencing without having to buy a $60,000 full-sized unit. Many desktop video systems are connected to two bearer channels of a BRI ISDN line. Two 64-Kbps lines are bonded together for a combined speed of 128 thousand bps. Often, these video systems are shipped ISDN-ready. The equipment needed to interface with an ISDN line, terminal adapters, are built into the video system.

If a video conference unit has a higher than 128-Kbps speed, users can bond together multiple BRI lines with aggregating devices. A common speed derived as a result of aggregating six bearer channels from three BRI lines is 384 Kbps. Connecting remote local area networks together is another common application, in addition to video, for aggregating devices.

ISDN for Data—The Need for ISDN at Both Ends

ISDN lines cannot transmit video or data to analog lines. ISDN circuits can only communicate with other ISDN-equipped services. Both BRI and PRI (ISDN with 24 channels) ISDN can communicate with each other. An engineering consulting firm that receives large graphics files is interested in receiving these files faster. However, they are concerned that their customers may not have ISDN. They are in the process of calling their clients to determine their clients' plans for ISDN. In this situation, BRI and PRI users can use ISDN for voice calls to end-users that have POTs.

BRI Pricing

Pricing for ISDN ranges. There are installation and monthly fees which range from telco to telco. Installation is in the range of $150 to $300. Monthly fees are anywhere from $10 to $40 over and above the residential charge for an analog line. Most telcos charge business customers extra for each minute of usage. Some plans for residential customers include 120 to 140 hours of data transmission. Others are flat rate, no charge for usage for residential customers. Business customers always have usage fees in conjunction with their ISDN service.

Primary Rate Interface ISDN—24 Channels

One difference between BRI and PRI is that BRI has three channels and PRI, primary rate interface, has 24 channels in the U.S. and Japan and 30 elsewhere in the world. Twenty-three are bearer channels for user data. Each bearer channel has a bandwidth of 64,000 bps. A twenty-fourth channel is used for signaling at the 64-Kbps rate. PRI lines connect either PBXs or multiplexers to telephone companies. PRI was initially available from interexchange carriers such as AT&T and MCI. It is now available from local telcos as well.

PRI lines are similar to T-1 because they have both have 24 channels. However, PRI ISDN has out-of-band signaling on the twenty-fourth channel. On T-1 circuits, the signaling is carried within each channel along with user data. The signaling capability enables the delivery of the calling party's telephone number, as described in using a PBX with a PRI line. On data communications, the signaling channel leaves each of the bearer channels "clear" capacity for all 64,000 bits. It does not use any capacity for the call setup or tear down signals. The twenty-fourth channel also tells the public network that the data calls should be sent over the public network's data network rather than its voice network. This is significant because all channels are available on-demand for voice or data. The data channels do not have to be reserved ahead of time strictly for video or data. The setup signals perform this function allowing more efficient utilization of the PRI trunk.

Whereas BRI ISDN runs directly from a user premise to a telco office, 24-channel PRI ISDN is installed on the "trunk" side of a PBX, or into a multiplexer. Trunk-side connections are run from central office to central office, or in the case of PRI ISDN, from the PBX to the central office. BRI ISDN can also be installed on PBXs as extensions. The BRI line is connected to specialized equipment within the PBX. Figure 6.6 illustrates a BRI line as an extension for a video conference unit. The PRI trunk runs from the telco to the PBX. When equipment, such as a video conferencing unit, attached to the BRI line dials a call, the call is programmed to go out on the PRI ISDN trunk. PRI ISDN service is for:

- Video conferencing at speeds generally from 64 Kbps to 384 Kbps.
- Sending the calling party number to large call centers.
- Backing up LAN to LAN connections.
- Backing up dedicated, private lines in case the private lines fail.
- Remotely accessing corporate and Internet service provider sites.

PBXs with PRI Trunks

PBXs are used with PRI lines for:

- Call centers, to receive the telephone numbers of callers.
- Video conferencing units that do not require the use of full-time ISDN service.

Large call centers use PRI ISDN to receive the telephone number of the person calling. With ISDN, the telephone number is sent at the same time as the call. However, it is

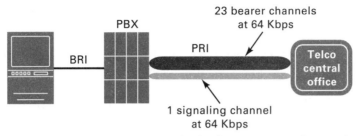

Figure 6.6   A PRI line for voice and video from a PBX to a telco switch.

sent on the separate D, or signaling channel. This is significant because it enables the telephone system, the PBX, to treat the telephone number information differently than the call. It can send the telephone number to a database that matches the telephone number to the customer account number and then it can send the customer account number to the agent's terminal that the call is sent to. It saves agent time on each call from having to type the customer account number into a PC. Having the telephone number called in a database also helps with demographic studies that measure the success of marketing campaigns and enhances analyzing the makeup of customers.

Some organizations use ISDN-compatible video conferencing systems as extensions of a PBX. This is called putting the video "behind" a PBX. In this way, the organization does not have to pay for dedicated BRI lines to its telephone company. Rather, the video equipment shares the PRI along with the voice telephone users. When the video is not in use, all of the PRI channels are available for voice. This capability of sharing the line for voice and data is due to the out-of-band signaling on the twenty-fourth channel. The signaling channel sends an identifier to the network telling the network that the video calls are data calls. The data call is sent on a separate part of the public network set aside for digital data calls.

PRI for Bandwidth-on-demand—Video and Data Devices Sharing a PRI Line

When a company has remote telecommuters with ISDN BRI at their homes for remote access to corporate email and databases, it frequently has PRI at its corporate locations. Telecommuters use BRI service to call into the corporate location's PRI trunk. Internet service providers that support ISDN access to the Internet also have PRI at their locations. They generally have multiplexers that handle multiple PRI trunks and T-1 to handle non-ISDN calls. In this way, they don't have to install separate devices for ISDN and non-ISDN calls.

Companies with multiple PRIs at a single location have the option of sharing one D signaling channel among multiple PRI trunks. For example, an Internet service provider with six PRI trunks might order two of them with D channels and the rest with 24 bearer channels to receive calls. They generally order more than one PRI trunk with a D signaling channel so that if one PRI with the signaling channel goes down, the second signaling channel provides the signaling for all of the PRI lines until the first signaling channel is restored.

ISDN multiplexers have what is called "bandwidth-on-demand" for applications such as video conferencing which require multiple bearer channels. For example, many companies want video systems capable of transmitting video at 384 Kbps, six bearer channels, because the video quality is superior to that of 128 Kbps. However, the video systems are not used 24 hours a day. With bandwidth-on-demand, when the video system is not in use, the six bearer channels can be used by other data applications. Bandwidth-on-demand provides an economic use of the PRI circuit. It allows it to be shared by devices that are not in constant use.

Digital Subscriber Line Technology

Digital subscriber line, DSL, technology was first introduced by Bellcore in 1989 as a way to send video and television signals from the telco central office to end-users over standard copper cable. At that time, video-on-demand was perceived as the broadband application that would drive digital subscriber line implementation. The first type of DSL technology was asymmetric digital subscriber line, ADSL, service. As its name suggests, ADSL is asymmetric. The line has different speeds away from and toward the customer. Larger video files can be sent "downstream" to the consumer. Less speed is used away from the customer, "upstream," to the central office to carry smaller files. The "upstream" portion needs less bandwidth because it is used to enter commands such as making a video selection.

Digital subscriber line technology was also seen as a vehicle with which to compete with cable TV companies. The Bell Operating Companies were looking for a vehicle to offer broadband, video and entertainment programming. An important consideration to Bellcore was that ADSL technology made use of the existing investment in copper cable. Fiber is not required to each customer's home or business.

The focus of DSL has broadened to include applications such as Internet access and work-at-home. Downloading entertainment and multimedia files is still on the agenda of DSL vendors. However, more of the focus on multimedia files is on downloading them from the Internet.

The increasing use of email, telecommuting and long Internet calls is creating traffic jams on the POTs network, which was designed for shorter voice calls. An advantage of DSL technology is its potential to relieve congestion caused by modem traffic on the public network. It does not send data traffic over the voice network. Digital subscriber line technology packetizes data traffic and sends it on a parallel data network. When DSL traffic hits the telco office, voice traffic is sent over the voice portion of the network and data traffic is routed on a separate data network. Modem traffic, on the other hand, is carried on the public network along with voice traffic. Modems convert digital data to analog so that local and long distance companies cannot distinguish it from voice calling.

There are multiple "flavors" of DSL technology. They run at different speeds, require different types of customer interface equipment, are envisioned for different types of customers and can run over variable lengths of copper cables. They are listed below and additional information is given in Table 6.5.

Table 6.5   DSL Speeds and Cable Requirements

Service	Upstream Data Rate	Downstream Data Rate	Top Distance from Central Office	Voice	Comments
ADSL	176 Kbps 640 Kbps	1.54 Kbps 6.14 megabits	18,000 feet 12,000 feet	Yes	In trials now at all Bells, GTE and MCI. Some availability in Pacific Bell and U.S. West territory.
HDSL	1.54 megabits	1.54 megabits	12,000 feet	No	Requires four wires; other DSL services only need two wires.
VDSL	640 Kbps 2.3 megabits	13 megabits 52 megabits	4,500 feet 1,000 feet	Yes	VDSL requires fiber optics on distances higher than these.
RADSL	176 Kbps 640 Kbps 128 Kbps	1.54 Kbps 6.14 megabits 600 Kbps	18,000 feet 12,000 feet 21,300 feet	Yes	This is a version of ADSL; the speed varies according to condition of the copper.
IDSL	128 Kbps	128 Kbps	18,000 feet	No	Limited availability with Internet service providers U.S. West and MFS.

- *ADSL:* asynchronous digital subscriber line. ADSL is asymmetric. The speeds on the line are different on data sent to the customer and on upstream data sent away from the customer.
- *HDSL:* high-bit-rate digital subscriber line. Not commercially available. Two competing standards on the market.
- *VDSL:* very high-bit-rate digital subscriber line. Not commercially available. Because of its high speed, it requires a combination of fiber and copper cabling.
- *RADSL:* rate adaptive digital subscriber line. A variation of ADSL that overcomes varying conditions and lengths of copper cable.
- *IDSL:* integrated services digital subscriber line. Works with the same customer equipment as ISDN. However, this is a dedicated service so customers do not have

to pay usage fees, just a fixed monthly charge with unlimited transmission. IDSL does not support voice calling.

Digital subscriber line services are being positioned:

* As a competitive service to cable modems.
* By competitive access providers on copper cabling that they resell from the Bell and independent telcos equipped with DSL gear.
* By interexchange carriers in partnership with CAPs or utility companies to share the cost of provisioning DSL.
* By Internet service providers over copper local loop facilities they resell from the Bell and independent telcos equipped with DSL gear.

DSL—For Internet Access and as a Response to Competition from Cable Companies

The main marketing thrust of DSL services is to compete against cable modems, ISDN and high-speed modems. The applications DSL is suitable for includes Internet access, access by telecommuters to corporate databases and high-bandwidth applications such as downloading graphics. There is particular interest in DSL service by RBOCs because it leverages their investment in copper cable in the outside cabling plant. Digital subscriber line services run on copper. Fiber to households and businesses is not required.

The Telecommunications Act of 1996 is a notable impetus in DSL deployment. The Act mandates the sale of separate network elements by Bell companies to competitors such as competitive local access providers, CAPs. CAPs can therefore purchase for resale the transport facilities, copper lines to individual businesses, and add high-speed DSL services to them. The CAPs do not have to run their own copper. They can use the Bell company copper lines for digital subscriber line implementation.

Obstacles to Digital Subscriber Line Availability— Cost and Ease of Implementation

Two stumbling blocks in DSL availability are the cost for exchange providers to install the service and ease of installation for end-users. Expense factors are the cost of the DSL modems themselves and the outlay for making the copper wires owned by local telcos suitable for DSL services. To be used for DSL, all loading coils and bridge taps must be removed from copper lines. (Loading coils boost the signal on analog copper telephone wires and bridge taps allow the same copper wire from the central office to feed multiple locations.) *Telecommunications* magazine, in its December 1996 issue, estimated that 20% of copper wires have loading coils.

DSL modems and gear are required both at the central office and at the end-user's premise. Initially, costs for the modems are in the $1500 range. Both *Network World*, January 6, 1997, and *Telecommunications* magazine, December 1996, reported that Bell companies want to pay $500 per modem.

In addition to affordability, ease of installation for end-users is a major issue. Business and residential consumers want a service that is easy to install and maintain. Availability of telephone support or having the DSL provider install and maintain the modems is something the industry is considering. One advantage of IDSL is that the service uses ISDN equipment that is readily available and that has become easier to install than when it was first available.

Frame Relay—A Shared Wide Area Network Service

Frame relay, first implemented in 1992, is a public network offering that allows customers to transmit data between multiple locations. By using frame relay, organizations do not have to plan, build and maintain their own duplicate paths to each of their sites. Consider a large retail chain. Without frame relay, it had over 100 dedicated lines, paths, for its own corporate use connecting all of its retail outlets to the corporate headquarters. It required a large investment in telecommunications staff to keep track of billing on each line, track repairs and sort out finger-pointing issues and plan capacity for each line. Finally, equipment at the headquarters location in the form of CSU/DSUs (digital modems, see Chapter 7) and multiplexers to terminate each of the over 100 lines was needed. These efforts resulted in large personnel, telephone company and equipment costs.

Frame relay is a network built by local and long distance telcos and shared by multiple users. It is a value-added, virtual, private network service. It acts like a private, dedicated network, but does not require end-users to lease their own dedicated lines. Frame relay service is an alternative to organizations building their own private data networks. It is a way to "leave the driving" to the network provider. Frame relay is used mainly for local area network to local area network data transmissions. It is generally cost-effective for organizations with more than four or five sites. Frame relay, though not without potential problems, has the following advantages:

- The network is managed by a long distance provider, not the end-user. This is critical for companies that want to concentrate on their main business, not on maintaining their networks.

- Less hardware is required at each location than that used for private, dedicated networks.

- Capacity on frame relay is more flexible than that of private lines. Many fast-growing small companies, such as high tech businesses, like the flexibility provided by frame relay to easily add capacity.

- Frame relay has its own internal backup routes so that customers do not have to provide multiple routes for reaching each location.

In contrast to dedicated, private networks, which are built and used for only one organization, frame relay is a shared network. Carriers such as local telcos and interexchange carriers build large frame relay networks shared by multiple organizations.

One reason why frame relay networks are faster than traditional packet switching networks is because they do not do extensive error checking in the network. User data travels at a high speed through the carrier's network. Older X.25 packet networks checked each packet many times as it traveled through the packet network. This slowed down the transmissions. With frame relay service, it is up to customers to perform error checking and checks for missing data within their own networks. Error checking and checks for lost packets is done by on-site routers. (See Chapter 1 for a description of routers.)

Connections to Frame Relay—Frame Relay Access Devices and Line Speeds

An important feature for organizations considering frame relay is the fact that individual sites do not need a direct path to every other site. Each site only needs a connection to the frame relay network. Each site that uses the frame relay services leases a circuit, a telephone line, from its equipment to a port on the frame relay switch. This line is called an access line. It provides access from user equipment to the frame relay network.

The access line itself can run at various speeds depending on the amount of traffic generated at each site. Sites at different locations can be configured with access lines at unlike speeds. Some frame relay vendors also offer dial-up, e.g., ISDN, access to their networks for customers with small sites. Dial-up services are also used as a backup in case the dedicated access lines to the frame relay network fails. Some of the options for access lines are:

- T-1—1.54 megabits.
- 56 Kbps.
- 128 Kbps.
- 256 Kbps.
- 384 Kbps.
- T-3—44 megabits.

Equipment on the customer premise end converts the traffic from the local area network into frames for compatibility with the frame relay network. This equipment is called a FRAD, or frame relay access device. It is often a card within the router. Each frame has bits telling the network when the user data (frame) starts and when it ends. These are the flags. There are also addressing and destination bits in the frame for billing and routing purposes so that the frame relay provider knows where to route and bill each frame.

Access lines transport user data to and from the frame relay network. Frame relay networks themselves pass the frames to high-speed switches to carry data from site to site within the carrier's frame relay network. The main technology the networks are run on is ATM, asynchronous transfer mode.

Frame Relay for Transmitting Voice

Organizations wishing to save long distance fees on voice traffic between company sites are starting to use frame relay to connect callers located at different sites (see Figure 6.7). They do not use frame relay for customer calls. While voice over frame relay is intelligible, it does not sound as good as the voice heard on the standard public network. Most of the traffic carried over frame relay is local area network to local area network traffic.

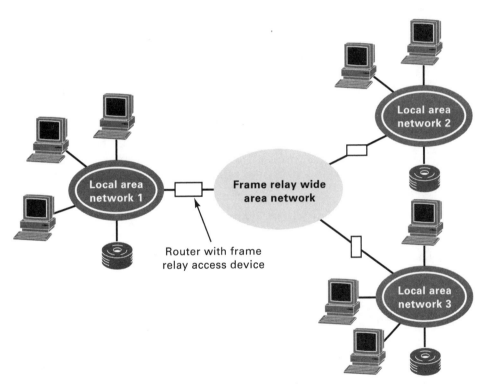

Figure 6.7 A local area network connection to frame relay.

The technologies used to transmit voice on frame relay are voice compression and silence suppression. Silence is suppressed so that pauses between words and the bandwidth used by one person listening are used to transmit data and voice from other users. In addition, the voice itself is compressed, made smaller, so that it does not require as much network capacity. Finally, voice traffic is given a high priority so that delay in voice conversations is minimized. There is a possibility, however, that if the frame relay network becomes highly congested, voice quality can decrease while delays between words increase.

Frame Relay Pricing—Ports, Circuits and Committed Information Rate

Frame relay pricing is a fixed monthly fee based on the following three elements, plus the cost of the telephone line used to connect each site to the frame relay service:

1. *The PVC, permanent virtual circuit*: A logical pre-defined path or link through a carrier's network. For example, if San Francisco and Tucson sites need to exchange data, the carrier defines a permanent virtual circuit between these two locations.
2. *The frame relay port*: The entry point, on a frame relay provider's switch, to the frame relay network. Multiple permanent virtual circuits can use one port. Ports are available in variable speeds such as T-1, 56 Kbps, 256 Kbps and 16 Kbps.
3. *The CIR, committed information rate*: The minimum guaranteed number of bps throughput, typically half the capacity of the port the customer is guaranteed to be able to send from each site. Some customers save money by using zero committed information rate. Customers can "burst," send data at the maximum speed of their frame relay port, even though the port speed is higher than the committed information rate. If the port is a 56-Kbps speed port, the location cannot send data through that port at higher than 56 Kbps speeds.

Potential Congestion on Frame Relay

Frame relay is a popular and growing network service. It is used frequently when organizations add new applications or build new networks. It saves them from leasing individual private lines, purchasing multiplexing equipment, designing the network and maintaining the network . However, frame relay is a value-added, shared network. Customers rely on carriers not overselling capacity. Once end-users have the service, they depend on their carrier's managing capacity, using the best telecommunications switches and providing them with reports on the success of their transmissions. If the carrier's network is oversubscribed, the carrier can drop frames. Organizations with mission-critical data communications or high levels of security requirements often hire the staff and spend the money to manage their own private networks.

ATM—Asynchronous Transfer Mode

ATM, asynchronous transfer mode, is a high-speed, currently up to 622 million bps, switching service capable of carrying voice, data, video and multimedia images. ATM, asynchronous transfer mode, is used mainly in telephone company networks. It is a higher cost service than frame relay for end-users to implement. However, large users are beginning to use ATM as a way to switch large files. For example, the entertainment industry is using it to ship film clips to other locations for editing. ATM is used by:

- Long distance providers.
- Bell telephone companies.
- Frame relay networks.
- Internet service providers.
- Large banking companies.
- Large universities.

ATM's speed is due to three characteristics:

1. The cells are fixed in size.
2. The cells are switched in hardware in a connection-oriented manner.
3. Switching is performed asynchronously.

Fixed-Sized Cells—Less Processing

Asynchronous transfer mode, ATM, packages the data it switches into discrete groups called cells. This is analogous to putting the same number of letters into each envelope. These envelopes are called fixed-sized cells. Handling fixed-sized cells requires less processing than switching with variable-sized cells. The ATM switch does not have to look for bits telling it when the cell is over. Each cell is a standard 53 bytes long. The switch knows when the cell ends. Five of the 53 bytes are header information. This includes bits that identify the type of information contained in the cell, e.g., voice, data or video, so that the cell can be prioritized. Voice and video, which need constant transmission (bit rate) so that there is no interruption in the voice or picture, need higher priorities than LAN data. Other header information is used for routing, putting the cells in the correct sequence and error checking. The remaining 48 bytes are the "payload," user data such as voice, video or sales proposals.

Switching in Hardware—Less Address Look-up

A significant reason why ATM is fast is that the cells are switched in the hardware. This means that an ATM switch does not have to look up each cell's address in software, which routers must do. Rather, an ATM switch sets up a route through the network when it sees the first cell of a transmission. It puts this information into hardware and the next time it sees a cell with the same header routing information, it sends it down the virtual path previously established. For example, all cells with XXX in the header use route 234. Using the same path for each cell makes ATM a connection-oriented service.

Asynchronous Switching—Improving Network Utilization

With asynchronous switching, every bit of the network capacity is available for every cell. This is different than synchronous multiplexing technologies such as T-1 and T-3. With T-3 multiplexing, every one of the 672 input transmissions is assigned a time slot. For example, terminal A may be assigned time slot 1 and terminal B assigned time slot 2. If terminal A has nothing to send, the time slot is sent through the network empty. ATM has no synchronous requirements. It statistically multiplexes cells onto the network path based on quality of service information in the header. For example, voice and video need better service, fewer delays and higher aggregate speeds than email messages. With ATM, this is accomplished without wasting network capacity.

Scalability—The Ability to Use ATM for Both High- and Low-Speed Applications

ATM can carry traffic of various speeds. It accepts streams from different inputs, e.g., telephone systems, routers and video devices, and sends them across paths, virtual circuits established by the ATM switch. This is scalability. ATM can be scaled from low-speed, 56 kilobit, to high-speed video and multimedia applications. Currently, ATM is installed mainly in carrier networks. As the technology matures, it is envisioned that more large corporations will use ATM as a way to carry both their voice and data traffic over wide area networks. For example, T-1 lines can run from a PBX to an ATM switch.

One way organizations upgrade to ATM switches is by upgrading their routers. Routers, which route calls between local area and wide area networks and between LAN segments, can be upgraded to ATM with the addition of network interface cards. LAN congestion is a prevalent problem. The use of Windows operating systems on PCs is one factor in causing traffic jams on corporate networks. Windows is a bandwidth hog. Other applications such as attachments on email messages, desktop video and sales proposals with graphics are additional high-bandwidth consumers. Currently, because of their expense, only very large customers are using ATM.

ATM switches are common in frame relay carrier networks for switching multiple customers' traffic at 622 megabits. ATM is faster than a T-3 multiplexer's speed of 45 million bps. Frame relay vendors use frame relay devices to connect directly to their customers. The frame relay traffic is passed from the frame relay port to ATM switches from which it is carried through the frame relay network.

Frame relay access to ATM switches located in frame relay carriers' networks is an interim technology for customers who want higher speed wide area network connections. The next step is to allow customers to access carriers directly from an ATM switch to an ATM port on the service provider's network. This will enable customers to access virtual private network services at faster speeds.

The most prevalent ATM speed is 155 megabits. The top current speed is 622 megabits. ATM switches are being developed which will run at Gigabit speeds, such as 10,000 megabits, which is equivalent to 10 Gigabits.

SONET—Synchronous Optical Network

SONET is a standard way to interconnect high-speed traffic from multiple vendors. Whereas ATM is a switching and multiplexing technique, SONET is a transport service used on fiber optic cabling. ATM is a Layer 2 service; it performs switching, addressing and error checking. SONET is a Layer 1 service. Layer 1 functions define interfaces to physical media such as copper and fiber optic cabling. SONET takes data and transports it at high speeds called OC, optical carrier, speeds. SONET links transport (carry) data from ATM switches, T-1 and T-3 multiplexers.

SONET is deployed in long distance and local telco networks. It was introduced in 1984 by Bell Communications Research Inc., the central research group jointly owned at that time by the RBOCs. The ITU has approved standards for OC speeds. The various OC levels and their speeds are shown in Table 6.6.

Table 6.6 Optical Carrier Levels

OC Level	Megabits	# 64-Kbps Channels	OC Level	Megabits	# 64-Kbps Channels
OC-1	52	672	OC-24	1,244	16,128
OC-3	155	2,016	OC-36	1,866	24,192
OC-9	466	6,048	OC-48	2,488	32,256
OC-12	622	8,064	OC-96	4,976	64,512
OC-18	933	12,096	OC-192	10,000	129,024

The number of channels in each optical carrier level is a multiple of the 672 channels in the OC-1 speed. An OC-3 line has $3 \times 672 = 2,016$.

SONET Rings—For Greater Reliably

The higher speeds attainable on fiber makes reliability extremely important. When a medium such as copper carries conversation from one telephone subscriber, a copper cut only impacts one customer. The SONET speed, OC-192, carries 129,000 transmissions. If the SONET ring that serves a major hospital, police department or armed forces unit fails, there can be an impact on the health, safety and possibly national defense of the locale. The FCC requires that telcos notify it if an outage affects more than 30,000 users. To add reliability, SONET deployment by telcos often uses ring topology. One set of fiber strands is used for sending and receiving; the other is a spare set. If one set of fiber strands is broken, the spare set reroutes traffic in the other direction. This is an important advantage over fiber run in a straight line. If there is a fiber cut on fiber running from one point to another, there is no other route for the traffic to take without the carrier intervening and rerouting calls. Figure 6.8 illustrates this concept.

Fiber Cuts Resulting from "Man-made" Conditions

Outages in telephone company networks are often caused by human error. Fiber optics and the equipment that enables it to transmit data at high speeds is reliable and has built-in duplicate hardware and software sections such as backup electricity and processors. The most outages occur when lines are cut. During hunting season, for example, hunters have shot at fiber lines, causing them to break. In the spring, digging equipment and back hoes break the fiber running underground. In the winter, ice on fibers causes them to break.

One company, Qwest Communications Corporation, based in Denver, Colorado, is building a SONET-based fiber network at OC-192 speed. It hopes to avoid these fiber cuts by burying the fiber it lays four and one-half to five feet deep. It is laying the fiber in 1/4-inch thick, high-density polyethylene conduit to make it impervious to environmental conditions. The only places it will not be buried is when it crosses one drawbridge and a railroad bridge. In these locations, it will be encased in a steel pipe.

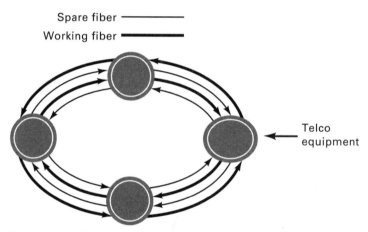

Figure 6.8   A fiber SONET ring where one pair of the fiber is used to transmit and receive data and the other pair is a spare for backup purposes. The backup fiber strands reroute traffic in the opposite direction of the fiber failure.

Telephone Company SONET Offerings

Local telcos are selling spare fiber-based SONET ring capacity to commercial customers. They are offering to interconnect T-1 and (mostly) T-3 services to the SONET rings for the extra reliability provided by the SONET ring structure. In particular, these services are aimed at call centers that require a high level of both capacity and, more important, reliability. Many call centers, such as airlines, which are in a competitive business, feel they lose large amounts of money in an outage because callers can call another airline if theirs is not available. The financial services industry is another industry that demands a high level of reliability.

One application for selling SONET transport provided by local telcos is to connect interexchange carriers' customers to the local telephone company's SONET rings. The speeds offered are at the OC-3, 155 megabits, OC-12, 622 megabits, and OC-48, 2,488 megabit, speeds. The local telcos claim 50-millisecond network restoral in the case of a network failure or degradation. They also run the SONET rings to multiple interexchange carrier switches and local telco facilities in the case of a failure at one central office. Matching SONET lightwave multiplexers are required at the customer premise and at the telephone company office to convert the SONET light signals to electrical signals, and vice versa.

SONET Technology—High-capacity Networks at Lower Costs

SONET speeds are increasing. This means less equipment and fewer fiber runs are required to carry vast amounts of traffic. This lowers the financial barriers for entry into construction of long distance networks. Qwest Communications Corporation is an example of this. Qwest is installing an OC-192, fiber-based SONET network. According to their Senior Vice President of Strategy and Planning, Qwest believes that motion, color and video applications will drive demand for network capacity. Qwest is using Nortel SONET gear and multiplexing eight OC-192 streams onto fiber. This will enable them to carry 80 megabits, or one million calls, on each fiber route.

New carriers are at an advantage in reaching OC-192 speeds. OC-192 requires special fiber called zero dispersion fiber. This fiber is thinner and has fewer impurities than standard single mode fiber previously purchased for carrier networks. New carriers do not have to contend with having lower grade fiber optic cabling in their cabling plant. Nor do they have to upgrade older multiplexers and SONET devices.

CHAPTER 7

Conversion Devices

This chapter takes the mystery out of connecting computers to telephone lines. Anyone sending data, video or images over a telephone line needs a conversion device between the telephone line and communicating data equipment. The conversion gear that sits between the communicating device and telecommunications jack is a DCE, data circuit-terminating device. When computers were initially used to transmit data over the public network, DCE devices were always modems because all of the telephone lines were analog and computers transmit digital bits. Modems convert a computer's digital signals to analog in such a way that they are compatible with analog telephone lines.

Transferring Data from Computers to Telephone Lines

Analog and digital telephone lines require different types of conversion devices. When digital telephone lines are used, DCE, data circuit-terminating devices, are required to modify digital computer signals to make them compatible with digital telephone lines. DCE devices for digital lines are called network terminals. NT1s are used for ISDN, integrated services digital network, lines and CSUs, channel service units, are used for other digital services such as T-1 and T-3 lines. Data circuit-terminating devices serve multiple functions. The functions they provide depend on the network services with which they are used.

Functions of DCE devices on digital lines include:

- Ensuring the correct number of 1s and 0s so that errors do not occur.
- Shaping the digital signal.

Functions of DCE devices on analog lines include:

• Converting digital computer signals to analog when data flows from computers to analog lines.
• Converting analog telco signals to digital signals when signals are received from the network.

Functions of DCE devices on both analog and digital lines include:

• Ensuring that data flows in an even, synchronous fashion by providing a timing source.
• Making sure the proper voltages are present.
• Performing error detection and error correction.
• Reducing distortion in transmitted signals.
• Compressing data so that more information can be transmitted.
• Providing a point up to which the network provider can test the telephone line so that problems can be diagnosed.

DCE gear with remote diagnostics can reduce the finger-pointing that is often present when maintenance staff try to determine whether repair problems are located in the network, computer, cables or modem. Problem determination, finger-pointing between computer suppliers, network vendors and modem suppliers, is a major problem with companies responsible for telecommunications networks. One way repair problems are pinpointed is by technicians sending test data bits to the DCE device. If the DCE device receives the data, the assumption is made that the problem is not in the telephone line or DCE.

When technical people talk about data networks, they often use the term DTE to mean the devices and multiplexers used for data communications. DTE stands for data terminal equipment. DTE gear includes:

• Laptop computers.
• Personal computers.
• PBXs.
• Video conferencing units.
• Key systems.
• Multiplexers.
• Facsimile machines.
• Automatic teller machines.
• Computerized cash registers.

Computers, PBXs, multiplexers, etc. communicate over a variety of local and long distance telephone services. These telephone services include the following:

- Analog POTs lines.
- Digital lines such as T-1 and T-3.
- ISDN.
- Analog cellular services.
- Digital wireless services such as PCS.
- Cable TV lines.
- Dedicated analog and digital lines (see Chapter 6).
- Access lines to frame relay networks.

Table 7.1 lists data circuit-terminating devices required to access particular types of network services. Compatible devices need to be at the sending and receiving ends of the communications channel. Standards have been set by the ITU, International Telecommunications Union, such that data communications equipment manufactured by multiple companies can speak to each other.

Table 7.1 DCE Devices as a Function of Telephone Lines

Network Service	DCE Devices
Analog lines, POTs	Modems
ISDN, integrated services digital network	NT1s
Other digital services, e.g., T-1 and T-3	CSU/DSUs, channel service units/data service units
Cellular	PCMCIA modem with cellular cable
PCS wireless	PCMCIA PCS-ready card
Cable TV lines	Cable modems

DCE—Direct Connections to Telephone Lines

All data circuit-terminating devices sit between the telephone line and computers or multiplexers, data terminating devices (see Figure 7.1). The increase in the computing power of chips and functionality of DSPs, digital signal processors, has resulted in shrinking the size of DCE devices. Modems used in laptop computers are about the size of credit cards. All DCE devices sit between the communicating device and telephone lines. However, a DCE device can be located in a slot of a computer. The following are sample configurations:

- Modems for PCs are either cards within a slot of the PC or stand-alone devices plugged into the back of the computer. In either case, they are plugged directly

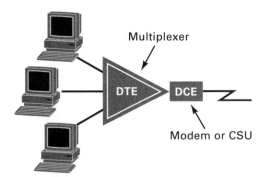

Figure 7.1   Multiplexers as data terminating
devices with a DCE.

into a telco jack. Telephones that share telephone lines with modems plug into the
back of the modem. Stand-alone modems need their own power source. Integrated
modems do not.

- NT1s and CSUs plug directly into a telephone company T-1 or ISDN jack.

- A BRI ISDN NT1 device can be an extension off a telephone system such as a
  PBX. In this case, the ISDN adapter plugs into a PBX jack the same way as a telephone. The PBX/key system is then connected to a PRI ISDN trunk.

- If a multiplexer is used, the multiplexer plugs into the DSU/CSU or modem, which
  is connected to the telco jack. Modems and DSU/CSUs can also be cards within
  multiplexers and connected by a cable to a telco jack.

- DSU/CSUs can also be cards within PBXs which act as multiplexers for digital
  services such as T-1 lines. A cable from the CSU/DSU card connects the PBX to
  the telco jack.

Modems—Analog Telephone Lines
with Digital Devices

Modems convert digital signals received from computers into analog signals to be
transmitted over analog lines. The process of converting analog signals to digital is called
modulation. At the receiving end, the modem demodulates the signal or converts it back
from analog to digital and transmits it to the DTE or computer.

When modems were first used, standards were set by the Bell System, in particular
AT&T, which controlled the types of devices that were allowed to connect directly to the
public network. (See Table 7.3 at the end of this chapter.) Starting in the 1980s, other
modem manufacturers began making higher speed modems. Standards began to be
approved by the Consultative Committee on International Telephony and Telegraphy
(CCITT), now known as the ITU, International Telecommunications Union. Rather than

wait for standards to be set, individual modem makers frequently developed and sold modems before the standards were set. This happened with V.fast, now the V.34 standard for 33.6 kilobit modems. Motorola and AT&T initially manufactured their own versions at 28.8 Kbps before the standard was finalized by the ITU.

To overcome noise and limitations on analog lines, modems perform error detection and error correction. Standards have been set for error correction such that modems from different manufacturers can correct errors in conjunction with each other. (See Table 7.3 at the end of this chapter.)

The major limitation of modems is the fact that they work in conjunction with analog lines. Circumstances that inhibit modem speeds are noise on lines and fading of the signal over distance. Even with error correction, modems do not consistently reach their optimal, stated speed. The higher the stated speed of an analog modem, the lower the percentage of time they actually transmit at that optimal speed. When the modem senses noise on the line, it decreases its speed. For example, a V.34 modem can step back from 33.6 Kbps to 19.2 Kbps. Many modems can increase their speed once line conditions improve.

Fax Modems

All facsimile machines have internal modems. The modem takes the image that is scanned into the fax machine as a digital image and changes it into an analog signal. When transmitting a fax message, the initial tones heard on the line are the sounds of the sending and receiving fax machines "shaking hands". The "handshake" consists of the sending and receiving modems agreeing on a speed and error correction method to use when transmitting the fax message. Fax machines connected to analog lines send at the ITU-specified Group 3 standard of 9600 bps.

Fax modems are also a feature of modems purchased as either cards or stand-alone units in PCs. Many owners of PCs now purchase fax/modems when they buy modems for their computers. Specialized communications software works in conjunction with their modem's fax capability such that documents they prepare in their word processing and spreadsheet programs can be faxed directly from their computers without having to be printed first and then sent from a stand-alone fax machine.

56 Kbps Modems—To Achieve Higher Speeds

In an effort to achieve higher data rates, modem manufacturers have developed a new modem technology aimed at providing faster speeds for the Internet access market. The 56 Kbps modems are *asymmetrical* in speed. The sending and receiving functions are accomplished at different speeds (illustrated in Figure 7.2). The send function, from the subscriber to the service provider, is slower than the receive function, from the Internet service provider to the subscriber. This difference is the result of the send end, the subscriber, having an analog line to the central office and the receive end, the Internet service provider, having a digital line. The subscriber, or analog end, is also known as the "upstream" portion of the transmission. The Internet service provider end is called the "downstream" portion. The assumption with 56 kilobit modems is that the Internet service

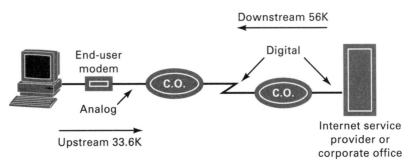

Figure 7.2   56 Kbps modems with faster speeds on the downstream digital portion of the transmission.

provider end is already digital, e.g., T-1 or PRI ISDN. Interestingly enough, many industry experts believe that this increase in access speed will result in more extensive use of the Internet. With slow-speed modems, Web pages take forever to download to subscribers who do not have the patience to wait for each page to be transmitted. Increasing the speed of downloading enhances user satisfaction with the service.

Competing Versions of 56K Modems

Just as the V.34 speed of 28.8 and later 33.6 modems was manufactured and shipped as V.fast before standards were agreed upon, the 56K modem standards are being shipped before standards are agreed to by the ITU. Interim standards will possibly be set by the U.S. Telecommunications Industry Association in late 1997. U.S. Robotics, now owned by 3Com, is shipping a 56K modem called the X2. Ascend, Motorola, Lucent and a consortium of manufacturers are shipping 56K modems using what they are calling K56flex technology. The K56flex modems use a computer chip manufactured by Rockwell International Corp. U.S. Robotics Corp. uses a chip from Texas Instruments Inc. Both types of 56K modems are shown in Table 7.2.

Table 7.2   The Two 56K Modem Types

56K Modem Type	Main Modem Manufacturers	Chip Used in Modem
X2	U.S. Robotics, part of 3Com	Texas Instruments Inc.
K56flex	Motorola, Inc., Ascend Communications and Lucent Technologies	Rockwell International Corp.

Modems made with the different chips cannot talk to each other at the higher speeds. Once the standards for 56K modems are set, many of the modems will be upgradeable with low-cost software upgrades.

Limitations of 56K Modems—Power and Attainable Speeds

Two problems, in addition to standards issues, exist with 56K modems. These are power requirements at the local telcos and conditions in the analog portion, the local loop part, of the public network. To achieve the 56 Kbps speed downstream, certain levels of power must be provided by the local telcos. If these levels are not present, the modems can only achieve 51- to 52-Kbps speeds. The FCC sets the power levels. No date is set for when power requirements might be changed.

The inconsistencies in the local loop, the portion of the telephone path from the end-user to the local telco, is another variable. Real-world performance does not consistently reach the standard speeds of 33.6 Kbps "upstream" and 56 Kbps "downstream."

A *Network World* article of January 13, 1997 reported that an anonymous modem manufacturer stated that its tests of V.34 modems, which have a top speed of 33.6K reached the 33.6 Kbps speed only 40% of the time. In any case, the vagaries of attempting to transmit digital data over analog local loops result in modem speeds. Meanwhile, modem manufacturers are working on the development of new modems to run at 114 thousand bps in the downstream direction. This speed also requires power modifications in the FCC specifications.

Modem and Fax Servers—Sharing Resources within Organizations

Modem and fax servers enable an organization's staff to share modems, fax machines and, more important, telephone lines. Before fax and modem servers became prevalent, individual users in organizations requested or went out and purchased, often without consulting the computer department, their own modems and desktop facsimile machines. This created maintenance, hardware and telephone expenses and confusion. It was not uncommon for a business with 400 employees to acquire 50 extra telephone lines to support individual modems. At $20 per line, the monthly expense for extra telephone lines was $1,000. The annual expense for these lines was $10,000. This did not include the cost of individual modems. In many cases, computer and telecommunications departments did not have a handle on where all of these lines were installed and who was using them when billing or repair problems arose. As more users wanted access to the Internet and email services, the problem of requests for individual telephone lines and modems escalated.

Not only did staff want to dial out of the building, they also needed to access corporate data files remotely from home or while traveling on business. Remote employees needed to access email, sales proposals, updates on shipping status and other corporate databases. Executives and telecommuters commonly needed to access large spreadsheet files when they traveled or worked at home. The most efficient way to handle both remote access into corporate files and requests for employees to transmit files from within buildings was to purchase central "banks" of communications devices and lines and make them accessible to employees.

For outgoing data communications, instead of giving each user his or her own modem, employees have access to modem servers located on the local area network. Any user with the appropriate security clearance located on the LAN has access to the pool of modems and fax services. The servers are cards located in PC slots. If more users than fax ports attempt to send files, buffers in the servers hold the messages until a port is freed up. Figure 7.3 shows this type of setup.

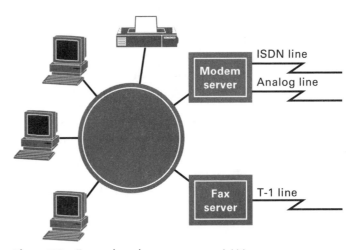

Figure 7.3   Fax and modem servers on a LAN.

These servers can be connected to a variety of telephone line types. For example, a single T-1 may be connected to a modem server so that 24 simultaneous messages can be transmitted. In addition, a mix of analog and digital telephone lines can be tied into a fax server; for example, two BRI ISDN lines and two analog lines. In this way, remote access for telecommuters and traveling employees can be supplied for people with and without ISDN.

PCMCIA Modems—Smaller Is Better

PCMCIA stands for Personal Computer Memory Card International Association. PCMCIA standards are for cards 3.37 inches long by 2.126 inches wide which plug into slots on portable computers such as laptop, palmtop and hand-held computers. They were initially designed as cards with extra memory for laptops. For example, if the hard drive on the laptop was too small, a PCMCIA card was installed to store extra documents or programs. These PCMCIA slots are now commonly used for modems and fax/modems.

PCMCIA modems can be used with:

• Analog POTs lines.
• Analog cellular services.

- Digital PCS services.
- Fixed wireless services.

PCMCIA modems are manufactured in a variety of speeds, including 56K. When plugged into a standard analog telephone line, they work the same way as standard, full-sized modems. An interface jack outlet for a telephone cord, as shown in Figure 7.4, is attached to the end of the card. Their small size is made possible by advances in silicon technology such that all of the modem's functionality can be put onto one chip. These functions include the digital signal processing, DSP, of noise cancellation and modulation and demodulation. (Changing the computer signal to analog and changing the received signals to digital.) Control services are added to the same chip as the signal processing functions. These control functions include error correction, compression and command functions such as dial the call, hangup, etc.

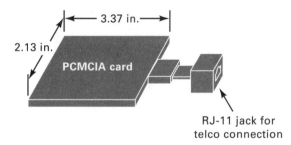

Figure 7.4 A PCMCIA modem card with a telephone line interface.

PCMCIA Modems with Wireless Services

When a PCMCIA modem is used to transmit data over cellular service, a special cable is required between the cellular telephone and PCMCIA modem. One end of the cable is plugged into the outlet on the PCMCIA card and the other is plugged into the cellular telephone. Manufacturers of cellular telephones such as Motorola, Oki and Radio Shack make proprietary cables. The Motorola cable, for example, has a three-wire bus with send and receive paths and a command path for commands such as go off hook, dial the call, etc. The modem also activates the speaker and microphone of the telephone so that progress tones are both audible and received. The cables cost between $50 and $100. (For a discussion of the limitations of using the analog cellular network for data communications, see Chapter 9.)

PCMCIA cards are also used for data communications over digital wireless PCS services. While called modems in the industry, they do not perform the modulation and demodulation functions of converting digital computer bits to analog signals and analog network bits to digital. In the case of sending digital computer bits over digital wireless

---

Portable Data Communications

According to Motorola, the two most popular uses of data communications over cellular networks are:

1. Email.

2. Facsimile transmissions.

Niche vertical markets of insurance adjusters and real estate agents frequently use cellular networks for data. For example, real estate agents download pictures of homes for sale onto portable, battery-operated printers. Customers can then decide on the spot if they are interested in viewing particular homes.

---

networks, the function of the PCMCIA card is to take computer-generated bits and convert them to a protocol the network will understand. However, standard communications software installed in portable computers work with these PCMCIA cards. The communications software uses the AT command set to set up the call. To this software, the PCMCIA card looks like a communications port used by DCE equipment.

Various digital PCS protocols, including GSM, TDMA and CDMA, are explained in Chapter 9. No single standard has been adopted in the U.S. GSM is used widely in Europe, Israel and Asia.

The digital PCS transmission techniques were designed with data communications in mind. Digital PCS networks, when used for data communications, do not have the high rate of errors and delays common with analog wireless services. More detail is provided in Chapter 9.

NT1s—Connecting Devices to an ISDN Line

DCE devices are needed for all data communicating devices, analog and digital. In most of the previous examples in this chapter, modems were reviewed as the DCE devices for analog, POTs lines. However, digital transmission techniques such as ISDN and T-1 services also require DCE gear. As reviewed in Chapter 6, ISDN enables voice, data and video to share one telephone circuit. On ISDN, the signaling is out-of-band. Dialing, ringing, busy signals, reorder tones, and dialed and calling telephone numbers are all carried in an out-of band channel.

Devices such as video teleconference units, PCs, PBXs, key systems and multiplexers that are connected to ISDN lines need an NT1 as the DCE interface to the ISDN line. The NT1 corrects the voltage on the signals and performs the Layer 1 functions as defined in the ITU standards. These are the electrical and physical terminations of the network. In addition, the NT1 provides a point from which line monitoring and maintenance functions can take place. The NT1 also changes the ISDN service from two wires that come into the building from the central office to four wires needed by ISDN equipment.

In the U.S., the FCC requires that the customer be responsible for supplying the NT1. In the rest of the world, telephone carriers supply the NT1.

While the NT1 plugs into the ISDN line the with one cable, another cable plugs into a TA, or terminal adapter. The TA does the multiplexing required on ISDN services. The multiplexing function enables one line to be used simultaneously for two voice or data calls plus the signaling, i.e., call setup, dialing, ringing, called number, etc., associated with the calls. ISDN telephones have built-in terminal adapters. TAs allow non-ISDN equipment to be connected to ISDN lines. Plain telephones, fax machines, video systems and computers can all use ISDN. They do, however, require an external TA if one is not included internally. Figure 7.5 illustrates a video conference system with an internal terminal adapter. The video system plugs into an NT1, which plugs into the ISDN line.

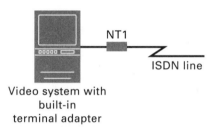

Figure 7.5 An ISDN line with an NT1 and ISDN-ready video conferencing system.

CSU/DSUs—Connecting Devices to a Digital Line

CSUs, channel service units, are required to interface with digital, non-ISDN telephone lines. These digital lines include: T-1, T-3, switched 56K and dedicated 56K lines. T-1 allows 24 voice, data and video transmissions to share one telephone circuit. T-3 circuits carry 672 voice, data and video transmissions on one circuit (see Chapter 6). Often, CSUs are cards within multiplexers and PBXs rather than stand-alone, external devices. Functionally, the CSU, sits between the telephone line and DSU, data service unit. The DSU connects to the data terminating equipment, DTE, such as the multiplexer. The CSU is similar in function to an NT1. It is a place from which maintenance and performance tests can be done. The CSU also provides clocking and signal reshaping. The clocking function is responsible for sending out bits in an evenly timed fashion. If the clocking is off, the transmission will not work. In this case, the technician might say, "the line is slipping," or "the timing is off."

The DSU makes sure the proper amounts of positive and negative voltages are present on the signals from the DTE to the CSU. Because CSU/DSUs are sold as one unit, many customers are not aware that there are actually two functions. At one time the DSU and CSU were sold as separate "boxes."

There are two main types of CSUs: super frame (sf) and extended super frame (esf). Extended super frame was developed by AT&T Bell Labs in 1985. It is superior to sf in its ability to be used to monitor T-1 performance while a line is in service. The type of CSU installed on the customer's premise should match the type used in the telephone provider's network. All interexchange service providers such as AT&T, MCI and Sprint have extended super frame capability. Local telcos support either super frame or extended super frame CSUs.

Cable Modems—Using Cable TV Facilities for Data Communications

Cable modems are used to provide high-speed connections from homes, schools and libraries to the Internet. They also, in some cases, provide municipalities with connectivity between government buildings. In exchange for granting franchises to cable companies to provide cable TV, city and town governments write into their agreements with the cable companies provisions whereby the cable TV supplier will allow free access of municipal departments to the cable plant for both Internet access and local area network connectivity. The local governments are required to pay for the cable modems and new equipment to work with the cable modems required at the cable company headend. A headend is the point from which programming is transmitted to local customers. The cable TV company agrees to make the cable plant suitable for data communications.

This suitability is the creation of "reverse" channels, or two-way capability. Television is currently a one-way broadcast medium. Television signals are transmitted from TV studios, via satellite, to microwave dishes located at cable operators' headends. From the headend, the television signal is transmitted via coaxial cable or a hybrid, a combination of fiber optic and coaxial cable outside the cabling system. Here is a sample cable TV franchise agreement in which the cable TV operator agrees to make the cable system capable of two-way communications by creating reverse channels from the subscribers to the headend:

> The Licensee shall construct a two-way residential cable television communications system available to all subscribers. The cable plant shall include the technical capacity for four reverse transmission channels for digital, audio or video transmission. The system will incorporate all necessary microwave reception equipment, satellite and terrestrial reception facilities, origination facilities and signal processing equipment at its headend.

The creation of reverse channels is done by using different frequencies for upstream and downstream transmissions. The upstream channels from the subscriber to the headend are in the 5 to 30 MHz or 5 to 42 MHz ranges. The downstream headend-to-subscriber channels are in the 54 to 350 MHz or 54 to 750 MHz ranges. Note that as in 56K modems, cable TV data communications are also *asymmetric.* The upstream and downstream portions are sent at different speeds. Splitting the frequencies into different ranges enables the cable to be used more efficiently. The same cable can be used for both sending and receiving. A separate cable does not have to be installed for the reverse channel capability.

Using cable TV for data communications is analogous to being on an Ethernet LAN. The Ethernet local area networking protocol is a shared protocol. All messages are broadcast onto the cable connecting devices on the LAN. There is no mechanism that ensures that everyone gets a turn to communicate. Local area networks need to be designed carefully without too many heavy users on each network. Ethernet protocols are used with TCP/IP. (TCP/IP is a suite of protocols that allow multiple networks to communicate together.) Like Ethernet LANs used within buildings, Ethernet used for cable systems' data communications is a shared medium. People using home computers for email and Internet access use the same cable facilities as people receiving television signals via CATV. Figure 7.6 illustrates how to set up cable TV to accommodate email and Internet access. (Some cable companies set up separate cable segments for municipal buildings in the towns in which they are located. These are called I-nets, for institutional cable.) Just as each PC on a LAN is a node on the network, each home PC user, library or school connected to the cable outside plant is a node on the Ethernet cable TV facilities.

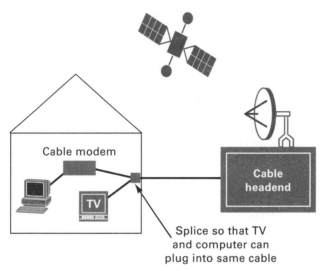

Figure 7.6   A hook-up of cable TV for data communications at a residential location.

When a home user connects his or her PC to the cable plant, he/she must purchase a cable modem from his/her cable provider. The modem plugs into either an Ethernet card in the PC or an existing Ethernet connection available on certain Mac computers. Another cable runs from the cable modem to an outlet that has been spliced into the cable provided by the cable company. TCP/IP software is required in the computer. Municipal buildings are connected to the cable plant with a combination bridge/modem unit. The bridge/modem plugs into the municipal building's LAN with one cord and into the outside cable connection with another cord. In both cases, a TCP/IP address is associated with each cable modem.

Infrastructures in Schools Needed before Internet Access

The Telecommunications Act of 1996 mandated subsidies for schools and libraries, to give them affordable access to the Internet. Internet access via cable TV systems already located within cities and towns is a viable avenue for this access. However, while the subsidy is helpful, it ignores the larger, more costly problem of cabling schools internally and upgrading schools' computers.

Providing what it takes in terms of cabling and upgraded computers in schools is a massive undertaking. The following is a sampling of what is needed:

1. Upgraded computers with sufficient memory and software for connections to LANs within buildings. Many school computers, when they exist, are old and limited. Computers are needed for libraries, individual classrooms, computer labs and administrative offices.
2. Cabling to link computers together. It is not enough to have stand-alone, upgraded computers. Each computer must be cabled to an in-building cabling system to access resources such as curriculum materials in a library and the Internet.
3. Hubs, patch panels, bridges, Ethernet cards in computers, servers and routers to connect the computers over the building wiring into a coherent LAN. An estimate from the California Educational Data Processing Association puts the cost for cabling and connection devices at $2800 per classroom.

Parent groups and businesses have banded together in some communities to build infrastructures in school cabling systems. An interesting example of this effort is "NetDay," started in California in 1996. The NetDay idea was originated by John Gage of Sun Microsystems and Michael Kaufman from KQED in San Francisco. It was eventually sponsored by 200 high technology companies and supported by the California Department of Education and thousands of volunteers. On March 9, 1996, thousands of volunteers wired 13,000 California schools.

The concept of NetDay has spread to other states, including Massachusetts, which had its first NetDay on October 19, 1996.

Cable modems provide the following functionalities:

- Equalization to compensate for signal distortion.
- Address filtering so that the modem only accepts messages intended for the correct recipient.
- Transmitting and receiving functions.
- Automatic power adjustments to compensate for power fluctuations.
- Adjustments in amplitude (signal strength or wave height) due to temperature changes.
- Modulation of the signal, i.e., analog to digital conversions, and vice versa.
- Compensation for delays caused by variable distances from the headend.

Items to keep in mind when using CATV technology for data communications include the following:

- Whether used for Internet access by residential customers or as a way to connect municipal buildings, each message is broadcast to every cable modem on the system. Only the modem with the specified address takes the message off the network. This may be a security issue to users considering using these services.

- Every message to another node is transmitted first to the headend and then to the addressee. With large files, there can be congestion on the network because each file transmitted goes to the headend and then downstream to each user until it is taken off the network by the addressee.

- Each modem requires local power.

- These are "nailed-up" connections where the service is always on. Residential users or people who work at home can leave their computers on all the time and be automatically notified when they have email messages. (A screen saver is used here so as not to burn out the PC's monitor.) Not only is this service many times faster than analog modems, but dialing into and logging onto the Internet, which can prove lengthy, are not required.

- An extra telephone line is not required for Internet access. However, a typical cable fee for unlimited Internet access is $50 per month. The savings might be $20 for a telephone line, plus $25 per month for a commercial internet provider such as AOL, which yields a total of $45 per month to offset the $50 for cable access.

- There is no agreed-upon standard for cable modems. The two main suppliers of cable modems are Zenith and Bay Networks Data Over Cable Division, formerly LANcity Corporation.

- While cable TV companies are starting to work out special agreements to municipalities for data communications and Internet access in return for cable franchises, cable for Internet access is not widely available to residential or general business consumers.

Modem Standards

Tables 7.3 and 7.4 list standard modem speeds for Bell and ITU. Standards have changed in two ways. Initially, they were set by the Bell System (AT&T). Now they are set by the ITU, the International Telecommunications Union headquartered in Geneva, Switzerland. In actuality, before the standards are agreed upon by the ITU, vendors ship new products they hope will become the new standard. This occurred with V.34 and with the 56K modems. Note that standard modem speeds have increased from 300 bps to 33,600 bps.

Table 7.3   Modem Speed Standards—From Bell 103 to V.34

Standard	Modem Type	Speed (bps)	Fall-Back Speed	Comments
Bell 103	Dial-up	300		
Bell 212	Dial-up	1200		
Bell 208A Bell 208B	Leased, dial-up	4800		
V.21	Dial-up	300	300	Used by Group 3 fax in half-duplex mode for negotiation and control
V.22	Dial-up, leased	1200	300	
V.22bis	Dial-up	2400	1200	Compatible with V.22
V.23	Dial-up	1200	600	Specifies 75 bps reverse channel
V.27ter	Dial-up	2400 4.8K	2400	Used by Group 3 fax for image transfer
V.29	Four-wire leased, two-wire leased, dial-up	9.6K	4.8K, 7.2K	Used by Group 3 fax for image transfer at 9.6K and 7.2K
V.32	Two-wire leased, dial-up	9.6K	4.8K	With V.42, provides 38.4K throughput
V.32bis	Two-wire leased, dial-up	14.4K	4.8K, 7.2K, 9.6K, V.32	Has rapid rate renegotiation feature for fast speed changes
V.32ter	Two-wire leased, dial-up	19.2K	18.8K, V.32 V.32 bits	An AT&T specification that is compatible with V.32 and V.32bis

Table 7.3 Modem Speed Standards—From Bell 103 to V.34 *(Continued)*

Standard	Modem Type	Speed (bps)	Fall-Back Speed	Comments
V.33	Four-wire leased	14.4K	12K	
V.34	Two- and four-wire leased, dial-up	33.6K	28.8, 19.2K V.32bis	Initial speed was set at 28.8K, known as V.fast prior to standard being set

bis = ITU term designating a second-generation standard.

ter = ITU term designating a third-generation standard.

V series standards, except for V.32, are promulgated by ITU.

Fall-back speed is the speed that a modem can drop down to with noisy lines.

Table 7.4 lists various error control standards and techniques available in modems. It also lists compression methods used. Compression removes white spaces and commonly repeated characters to improve throughput (efficiency) on data communicated. The repeated characters are replaced by abbreviated versions of the characters. A text file sent using a 4:1 compression scheme may only need to send 25% as many bits as a non-compressed file.

Table 7.4 Modem Error Control and Compression Standards

Standard/Technique	Description
V.42	Specifies both Microcom Networking Protocol (MNP) 2-4 and Link Access Procedure (LAP) M error correction for full duplex modems. Compatible with V.34, V.22, V.22bis, V.26ter, V.32, V.32bis and the proprietary versions of V.32 ter.
V.42bis	A data compression protocol that theoretically allows up to 4:1 file compression.
V.54	A standard for local and remote loop test devices in modems.
LAP M	The preferred error control protocol specified by the V.42 standard for error control in modems.
MNP 1	MNP 1 has been superseded by later versions.
MNP 2-4	Part of the V.42 standard for error control.

Table 7.4   Modem Error Control and Compression Standards *(Continued)*

Standard/Technique	Description
MNP 5	A data compression protocol that provides a 2:1 compression ratio.
MNP 10	Optimizes modem performance over adverse line conditions and cellular links.
EC2	Developed by Motorola. Modifies cellular transmission levels to avoid clipping, which is common at high frequencies.
TX-Cel	An error correcting algorithm that reduces cellular amplitude distortion.
ETC	Enhanced Throughput Cellular (ETC), newer than MNP 10, works with V.42 bis, the ITU-specified standard error control that works with V.42bis. Enhances throughput on cellular.
ETC2	AT&T's enhanced version of ETC. ETC2 will be used in conjunction with AT&T's proposed fixed cellular service. Enables cellular modems to operate at V.34 speeds of 21.6 Kbps from a stationary position.

CHAPTER 8

The Internet

According to a poll commissioned by *Business Week*, conducted by Baruch College and Louis Harris Associates April 11–16, 1997 and published in its May 5, 1997 edition: "The *number* of Americans using the Net has nearly *doubled*, to around *40 million* adults, over the past year." *Business Week* concluded that the Internet is changing the way American industry conducts business. Businesspeople assume that customers, suppliers and partners have the ability to communicate electronically over the Internet. Residential consumers commonly use the Internet to exchange electronic mail with family members and for research, and children and adults spend time on the Internet playing games and using entertainment services.

The Internet is a connection of multiple networks. The networks communicate with each other over a suite of standardized protocols, TCP/IP, which sends data over the Internet broken up into "envelopes" of data called packets. Internet traffic is sent over many of the types of telecommunications switches and lines such as ATM and T-3 discussed in Chapter 7. The high-speed lines are the backbone of the Internet. They carry the highest amount of Internet traffic. The Internet backbone transmits requests for information and the information itself on the Internet.

The World Wide Web was created to make "surfing" the Internet easier. It is a way to link documents such as home pages on distant computers together. Users can point and click their way from computer to computer on the Internet. Before the World Wide Web was developed, documents on the Internet were available only as text. There were no pictures, no "buttons" to click on to issue commands and no advertising banners. There was also no color. Everything was black and white. The World Wide Web is not separate from the Internet. It is a way to navigate from resource to resource on the Internet by clicking on highlighted text or "buttons" on screens. It is additionally a "multimedia" way to view information on the Internet. It makes information available in the form of graphics, sound, text and video.

Access to the World Wide Web is done via software called browsers. Menu-driven browsers such as Internet Explorer, Netscape Communicator and Netscape Navigator offer computer users a "menu" of choices in plain English. For example, pull-down menus or graphical "buttons" provide an easy way for users to tell the computer to print a file, go back to the last Internet "site" visited, delete a file or find information.

The History of the Internet

The Internet was started in 1969 by the Department of Defense's Advanced Research Projects Agency, ARPANET, in a computer room at the University of California, Los Angeles. The point of the Internet was to enable scientists at universities to share research information created in distant locations. ARPANET was created 12 years after Sputnik, during the "cold war." The original goal of ARPANET was to develop a network secure enough to withstand a nuclear attack. According to an August 24, 1994 article in *Network World*, Leonard Kleinrock stated, "ARPANET was conceived as a way to save money by getting government-funded researchers to share computers rather than each of them buying their own."

The first communications switch used to route messages on the ARPANET was developed at Bolt Beranek and Newman, BBN, in Cambridge, Massachusetts. ARPANET was based on packet switching developed by Rand Corporation in 1962. Data was broken up into "envelopes" of information that contain addressing, error checking and user data. One advantage of packet switching is that packets from multiple computers can share the same data line. A separate path is not needed for each transmission. Moreover, in the case of an attack, if one computer goes down, data can be rerouted to other computers in the packet network. TCP/IP, the protocol still used on the Internet, was developed in 1974. It supports a suite of services such as email, file transfer and logging onto remote computers.

BBN and Construction of the Internet

Bolt Beranek and Newman, now called BBN, is the company that built ARPANET, the original Internet network. BBN, now owned by GTE, started out in Cambridge, Massachusetts in 1948 as an acoustics company. According to an article in *The Boston Globe* on May 7, 1997, BBN designed the acoustics for the hall in the United Nations where the General Assembly meets. In the late 1950s, Bolt Beranek and Newman hired a psychologist named J.C.R. Licklider. Licklider pointed BBN in the direction of computer research.

Licklider saw the potential of computers and began hiring computer specialists from Harvard and MIT. *The Boston Globe's* May 7, 1997 article quotes Katie Hafner's book *Where Wizards Stay Up Late* on the history of the Internet, "BBN was thought of as sort of the third university. People would go from Harvard to MIT to BBN." Licklider went from BBN to the Pentagon. From there he headed the ARPANET project. BBN was contracted to build the first ARPANET network connecting four universities: UCLA, UC Santa Barbara, University of Utah and the Stanford Research Institute.

In 1984, as more sites were added to ARPANET, the term Internet started to be used. The ARPANET was shut down in 1984 but the Internet was left intact. According to the June 1997 issue of *Telecommunications*, by 1984 there were 500 computers on the Internet. In 1987, oversight of the Internet was transferred from the Department of Defense to the National Science Foundation.

While still used largely by universities and technical organizations, applications on the Internet expanded from its original defense work. In particular, newsgroups, also called alt.groups, proliferated. These groups, used by computer hobbyists, college faculty and students, were formed around special interests such as cooking, specialized technology and lifestyles. The lifestyles newsgroups included sexual preferences (gay and lesbian), religion and gender issues. Computer-literate people were also using the Internet to log onto computers at distant universities for research and to send electronic mail.

The Internet was completely text prior to 1990. There were no graphics, pictures or color. All tasks were done without the "point and click" assistance of browsers such as Netscape and Internet Explorer. Rather, people had to learn, for example, UNIX commands. UNIX is a computer language developed in 1972 by Bell Labs. UNIX commands include: *m* for get mail, *j* for go to the next mail message, *d* for delete mail and *u* for undelete mail. The Internet was not for the timid or computer neophytes.

The advent of the World Wide Web in 1989 and browsers in 1993 completely changed the Internet. The World Wide Web is a graphics-based vehicle to link users to sources of information. It is based on a method whereby users "click" on items of text to be transferred to a site with information on the highlighted text. In 1993, the Mosaic browser was developed at the University of Illinois as a point and click way to access the World Wide Web. This opened up the Internet to users without computer skills. It is no longer necessary to learn arcane commands to open mail, to navigate from site to site for research or to join chat or newsgroups.

In 1995, the National Science Foundation turned the management of the Internet backbone over to commercial organizations. Commercial networks such as MCI, Sprint, UUNET (now part of LDDS WorldCom) and BBN carry a large portion of the backbone Internet traffic.

Bulletin Board Systems (BBSs)

Bulletin boards are used independently from the Internet. They allow people with modems connected to their computers to read information and post information on a PC. The following are examples of how bulletin boards are used:

- Information on bugs in software.
- Features of new software.
- Downloading engineering designs.
- MCI rates and services for telecommunications consultants.
- Information on specifications for requests for proposals.

- Downloading new software releases.

- Computer society notices.

Users throughout the 1980s used modems, personal computers, communications software and telephone lines to dial into information on other computers. These information services are called bulletin boards. Bulletin boards are another way, independent of the Internet, for users to exchange information electronically. Many of them were used to "chat" and exchange ideas around specific hobbies. For example, callers would dial in and type ideas or experiences they had with new software or computer equipment.

These bulletin boards, many of which still exist, are made up of modems and telephone lines connected to PCs. One application is the dispersion of technical information. For example, technicians might dial into a bulletin board for information on the latest software release of a piece of equipment. Meeting notices were also posted on bulletin boards. Often, these BBSs, bulletin board systems, were operated by computer hobbyists around a common interest. Operators of bulletin boards are called SYSOPs, short for system operators.

Although many bulletin boards still exist independently of the Internet, they are being replaced by sites on the World Wide Web. Instead of having to learn a telephone number and logon procedures, people can "go to" World Wide Web locations if they know a Web address. An example of an organization now on the World Wide Web that might have previously been on a bulletin board is a group of stamp collectors. Stamp Collectors Exchange has a Web site where collectors can list stamps they want to buy or trade.

Who Runs the Internet?

The Internet is run informally by a number of organizations. The key ones are:

- *The Internet Society, ISOC:* ISOC is a nonprofit group that promulgates policies and promotes the global connectivity of the Internet. The group is the closest thing to a governing body for the Internet. It was formed in 1992 and is open to anyone who wishes to join.

- *Internet Engineering Task Force, IETF:* IETF is a standards setting body. The IETF works under the aegis of the Internet Society. It focuses on TCP/IP protocol standards issues. TCP/IP is the protocol used on the Internet.

- *Network Solutions Inc.:* Network Solutions Inc. was given the task by the National Science Foundation in January of 1993 to register Internet names for organizations that want new addresses. The registration service is called the Internic, Internet Network Information Center.

- *IOPS.ORG, Internet Operators' Providers Services:* IOPS was formed in May of 1997 to address Internet routing robustness—where to send packets based on conditions such as congestion. It was founded by nine of the largest Internet service providers, including AT&T, GTE and MCI. The point is to establish standard procedures on routing data between multiple operators' networks.

- *The World Wide Web Consortium:* The World Wide Web Consortium, also known as W3C, is a group formed to develop common standards for the World Wide Web. It is run jointly by the MIT Laboratory for Computer Science, the National Institute for Research in Computer Science and Automation in France, responsible for Europe, and Keio University in Japan, responsible for Asia. Over 150 organizations are members.
- *Internet Assigned Numbers Authority, IANA:* IANA coordinates setting standards for Internet addressing. Suggestions have been made by them for ways to increase the number of available Internet addresses. A proposal has been made by the IANA to create an additional 27 organizations, in addition to Network Solutions Inc., to register Internet addresses.

Who Owns the Internet?

No one organization owns the Internet. Rather, the Internet is a worldwide arrangement of interconnected networks. Network service providers, including AT&T, Digex, Sprint, Metropolitan Fiber Systems, UUNET and MCI, carry Internet information such as email messages and research conducted on the Internet. These networks are worldwide in scope with backbone networks run by network providers in countries such as France, Germany, Japan and others that have Internet networks.

Commercial organizations that own the high-speed lines that make up the Internet transfer data between interconnected networks at locations called "peering" sites. At the peering sites, network devices called routers transfer messages between the backbone, high-capacity telephone lines owned by dozens of network service providers.

In 1995, the National Science Foundation funded four of these "peering," or network access points. They are located in New Jersey, Washington, DC, Chicago and San Francisco. These sites are now run by commercial organizations. MFS (Metropolitan Fiber Systems), an LDDS company, runs MAE East, one of the sites started by the National Science Foundation. MAE stands for metropolitan-area exchange.

In response to concerns about traffic at these peering centers "bogging" down the Internet, private peering exchanges are set up by network service providers such as BBN, MCI, Sprint and PSINet. "Meeting" places to exchange data have been set up to avoid possible congestion at the major exchange centers. This direct exchange method is seen as a more efficient way to exchange data. Moreover, carriers agree on levels of service, amount of data to be transferred and delay parameters. They feel they can monitor reliability more closely at private peering exchanges.

Internet Services

Prior to 1995 and the availability of the World Wide Web and browsers, using the Internet and sending email was done without menu-driven software. PC owners had to know the "commands" to use the Internet. People who surfed the Internet did so via services such as FTP, file transfer protocol, and Telnet. They sent and received electronic mail through a service called SMTP, simple mail transfer protocol. All of these services relied on users knowing the commands of a computer language. Faculty and students who accessed the Internet from their colleges had to either obtain manuals of instructions put together by their computer information staff, share tips on using the Internet from local computer societies or purchase books such as *Zen And The Art Of The Internet*.

Researchers used file transfer protocol, FTP, to log onto computers at other sites, such as other universities, to retrieve copies of text files. Files were in text form. Graphics, video and voice files were not available. Moreover, finding information was a complex task. Researchers were able to search at thousands of sites worldwide. Commands had to be typed into the computer in an exact format. Dots, spaces and capitalization rules were strict. For example, "dir" let you see the contents of a directory, while "get filename" let you view the file on the computer screen. To simplify the search process, programs such as Archie were created. Archie was meant to simplify FTP use by enabling searches by topic. Gopher, introduced in 1991 by the University of Minnesota, a precursor to Web browsers, was more menu-driven than Archie and Veronica, but was bypassed, after 1994, when Mosaic, an early World Wide Web browser, was developed.

Another service available to access information prior to the availability of browsers is Telnet. While file transfer protocol is a way to transfer a file, Telnet is an Internet service for creating an interactive session with a computer on a different network. It lets users log onto another computer located on the Internet as if they were a local terminal. People used Telnet with arcane commands such as "host name" prior to the availability of browsers. They had to know the name of the remote computer they wished to log onto. Telnet and FTP are still used; however, access to them is via menu-driven software called browsers, e.g., Netscape Navigator and Internet Explorer.

The World Wide Web—Linking and Graphics

The World Wide Web was conceived as a way to make using and navigating the Internet easier. It is not a separate part of the Internet. It is the graphical way to use the Internet. The World Wide Web enables users to hear sound and see color, video and graphical representations of information. Moreover, it provides links to information using text and graphic images embedded in documents to "navigate" to other Web sites. These links are in the form of highlighted text and graphics. Users click on them with a mouse to move from one document to another or from one site to another. These two capabilities, linking and graphics, are the strengths of the World Wide Web.

People can now use the Internet without knowing arcane computer commands. They don't have messages on the bottom of their screens with "command" lines where they need to type in computer commands such as "top" to return to the top of a Main Menu. Moreover, multimedia capabilities of voice, full motion and graphics create an entertainment and advertising medium.

The World Wide Web was created in 1989 by Tim Berners-Lee at CERN, the European Laboratory for Particle Physics. The goal of creating the Web was to merge the techniques of client server networking and hypertext to make it easy to find information worldwide. The basic concept is that any type of client, the PC, should be able to find information worldwide without needing to know a particular computer language or without needing a particular type of terminal. Access should be universal.

The name of the protocol used to link sites is hypertext transfer protocol, HTTP. HTTP are the letters that start Web addresses. When a browser sees HTTP, it knows that this is an address for linking to another site.

Hypertext Markup Language (HTML)—Web Speak

Hypertext markup language, HTML, made available in 1991, is the tool that Web page creators use to write Web documents. Web pages are documents created with HTML. HTML is the authoring software that controls the "look" of a Web page. Employees who write Web pages for their organizations' home pages use HTML commands. Each hypertext command begins and ends with the <> signs. For example, bolded text is prefaced with <b>. Early users had to know the commands themselves. New HTML word processing software used for creating Web pages has embedded HTML commands.

The ease of writing home pages has added to their proliferation on the World Wide Web. Individuals, businesses and non-profit organizations can create home pages. This is a factor in the increasing number of sites on the World Wide Web.

Home Pages

A Home page is the default first page of a World Wide Web site that users see when they visit an organization's or individual's Web site. They are documents created with Web authoring software, hypertext markup language. All Web pages are linked together with hypertext links. These are the highlighted pieces of text that transfer surfers to another site or page when a user clicks on them with a computer mouse.

Maintaining a Web site is an ongoing task. Often, the sites that hypertext links connect users to become defunct or change their address. This results in readers clicking on highlighted text that goes nowhere. Some non-profit organizations use their Web sites to post meeting notices and membership lists. In these cases, meeting notices must be updated. Membership lists are moving targets. In particular, members' email addresses and telephone numbers change frequently. Keeping Web sites current and accurate is a major task.

Home pages are provided by:

- *Information service providers:* Prodigy, America Online and Microsoft Network.
- *Major corporations:* Most major corporations have home pages located either on servers on their own premises or on the premises of network operators such as MCI, AT&T or Sprint. These organizations are "hosts" for company home pages.
- *Individuals:* Many individuals use their information service providers or Internet service providers to host their Web pages.
- *On-line newspapers and journals: The Wall Street Journal* and many other newspapers such as *The New York Times* offer on-line editions.

• *Universities, schools, hospitals and towns:* These organizations all offer home pages. In elementary schools, some classes have started their own home pages so parents can keep up with classroom events. Non-profits use the same types of hosting services as major corporations.

Hosting—Computers Connected to the Internet with Home Pages

A World Wide Web host is the computer at which documents or databases that users read are located. The World Wide Web was designed without any centralized facility. Anyone who wishes to may make information available to the Internet community. They can "publish" documents on the Web. It is based on a client server model. The client is the device, usually a PC, that reads information located on a server, called a host. The server is a computer sized to the requirements of the application, or documents it is "hosting".

Hosts are located at large universities, corporations, cities and towns and Internet service providers. Because of the cost of providing both the telephone connections to the Internet and the security to keep hackers out of non-internet computer files, hosting is often done at the Internet service provider site. Internet service providers, hopefully, have enough telephone lines to cope with surges in volumes on popular Web sites. When organizations plan home pages, they are often not certain of the volume of people who will access them. Therefore, they often initially put them on Internet providers' servers. Large Internet service providers are more likely to be able to keep up with uncertain volumes for new home pages.

Browsers—Moving from Web Site to Web Site

Browsers are the graphical interfaces between users and the World Wide Web. They provide the graphics capability associated with the Web. Browsers make the Internet "friendly" and easy to use. Without browsers, users had to be knowledgeable enough to use computer commands to find sites on the Internet. With browsers, people can go directly to an organization's home page with the help of bookmarks. Bookmarks are a speed-dial list of sites visited often. Addresses saved as bookmarks can be accessed by merely highlighting the bookmark and hitting the return key. In addition to bookmarks, browsers have "buttons" that when clicked transfer Web surfers to different sites and to advertisers .

The first browser, Mosaic, was created by the National Center for Supercomputing Applications at the University of Illinois and Europe's CERN (the European Laboratory for Particle Physics) research laboratory. The University of Illinois did not have the staff to support commercial help with the browser and gave the license to Spyglass, Inc. to commercialize. In return, The National Center for Supercomputing Applications expected Spyglass to pay it royalties from sales of Mosaic. However, Mosaic was eclipsed, with the help of Mosaic developers who moved to Netscape Corporation, by Netscape Navigator.

Netscape Navigator is sold by Netscape Communications. Netscape Communications, originally called Mosaic Communications Corporation, was started in 1994 by Jim Clark of Silicon Graphics and people who had created Mosaic at the University of Illinois.

In six months, the Netscape team had created a browser they referred to in-house as Mozilla, the Mosaic-Killer. Netscape, however, was developed not to be compatible with Mosaic. Mosaic was created by government funding to be a standard Web browser.

Netscape's original strategy was to make its Netscape Navigator free when downloaded from the Internet. The idea was to set its own de facto standard by flooding the market with Netscape Navigator. Netscape envisioned selling Web page design tools later that would cost from $1500 to $50,000. These server software tools are now sold for Internets, intranets and extranets. (See below for intranets and extranets.) Beta and trial versions of Netscape Navigator and new browsers are still free on the Internet. They do charge for final versions of their browsers. Netscape Communications also spawned sales of software programs called plug-ins to work with its browser.

The development of Netscape Navigator was a significant step away from government standards and toward commercialization of the Internet. The following is a quote from Nathan Newman's February 1997 on-line newsletter *Enode*. The quote is a response by Jim Clark, the founder of Netscape Communications:

> At some level, standards certainly play a role, but the real issue is that there is a set of people, a set of very powerful companies out there, who don't play the standards game....Companies such as Microsoft aren't going to sit around and wait for some standards body to tell them, you can do this. If your philosophy is to adhere to the standards, the guy who just does the de facto thing that serves the market need instantly has got an advantage.

Netscape became the fastest growing software company in history. It made $81 million in its first year of operations. According to *Business Week*'s February 10, 1997 issue, Netscape Corporation's 1997 revenue is expected to reach $500 million. Netscape browsers are also used on local area networks in many commercial organizations for individual employees' PCs. The employees use Netscape as their interface when they access the World Wide Web. It is installed on all new Mac computers.

Meanwhile, not to be outdone, Microsoft introduced its own browser, Internet Explorer, in 1995. In an attempt to control the browser market, Microsoft installs Internet Explorer on all of its new Windows PCs. As a matter of fact, according to the *Business Week* February 10, 1997 issue, virtually every new PC comes with Internet Explorer installed as the browser. In the same issue of *Business Week*, Bill Gates is quoted as saying, "The Internet is the most important thing going on for us. It's driving everything at Microsoft."

Email—Computers to Send, Store and Receive Messages

Electronic mail is the computer-based storage and forwarding of text-based messages. Electronic mail, email, was invented in 1972 by an engineer at BBN in Cambridge, Massachusetts, Ray Tomlinson. BBN, then Bolt Beranek and Newman had built the first Internet network, ARPANET, and users of ARPANET needed a way to communicate with each other. Standard electronic mail is based on a service that is part of the suite of protocols called TCP/IP. This service is called SMTP, simple mail transfer protocol.

When electronic mail is sent through the Internet, it is sent using the SMTP protocol. This protocol specifies addressing conventions, the @ sign with user names and locations, ways to send copies to other people and the fact that the characters are sent in ASCII, American Standard Code for Information Interchange. ASCII, a computer code for translating computer bits into characters, has a limited number of characters. It supports all of the upper- and lower-case letters of the alphabet, numbers and symbols such as *, $, underlining and %. ASCII, however, does not include formatting options such as italics, bolding and columns. If a word processing file is sent as an email message, it loses its formatting.

Electronic mail is a service independent of the Internet. Commercial messaging services such as MCIMail, Sprint Mail and AT&T Mail offer email without Internet access. Before they had connections to the World Wide Web, organizations signed up for these commercial electronic mail services as one way to exchange electronic mail with customers or remote offices.

Electronic mail is an important vehicle for organizations to send messages internally. Once establishments connected their computers together in local area networks, they started using electronic mail to communicate with each other within the same buildings. The next step in connectivity and messaging was to tie each location's electronic mail and computers together in wide area networks, WANs.

Once firms had internal email, they extended their use of email to communicating with business partners in outside companies via the Internet. Businesses now have direct telephone lines from their locations to the Internet. They use email to check the status of projects, follow up on proposals and check the status of orders. In my own teaching at Northeastern, students send me email messages with questions about technical issues, exams and attendance. Email is faster and less intrusive than a telephone call. A telephone call is an interruption, whereas email is stored and can be reviewed when convenient. Users can pick up their email at scheduled times during the day. Other people like email because, unlike voice mail, it provides a hard copy record of communication.

Email is so widespread within businesses that staff are sometimes bombarded with messages. One computer manager found 300 messages in her mailbox when she returned from a two-week vacation in Italy. Sorting through lengthy messages is a challenge. Managers now have voice mail, fax messages and email to review. Information overload is a common complaint heard from people who work in large firms where broadcast messages on a variety of topics are common. Departments commonly send broadcast messages, the same message to many people, on topics such as the next blood drive, benefits, network shutdowns, lost and found items, new products and new internal telephone numbers.

Email is changing the way both commercial and residential consumers communicate. The June 16, 1997 "Technology" section of *The Wall Street Journal* cited Forrester Research Inc. in Cambridge, Massachusetts statistics that one-fifth of the population has email capability. Much email is between people in the same family. Because of the proliferation of email and computers at college campuses, students and parents often stay in touch via email. The SOHO, small office home office, is another example of the use of home computers for email. Once someone has a computer to work at home, the next logical step is to join the World Wide Web via an Internet or information service provider such as AOL, CompuServe or a local Internet service provider.

Email Attachments—To Aid Collaborative Projects

Firms now use their email services to exchange more than ASCII text messages. As companies use consultants and business partners on joint projects, they often need to exchange spreadsheet and word processing documents. For example, they may want feedback on work they have done or they may each work on separate pieces of the same project. Faxing is limited for large documents and overnight mail is expensive. Each of these methods, moreover, wastes paper and requires that the document be re-typed into a computer. One standard for sending documents is MIME, multi-purpose Internet mail extensions.

MIME includes a standard way to attach bits at the beginning and end of the attachment telling the receiving computer what type of file is attached and when the attachment ends. For example, the bits may tell the computer, this is a Microsoft Word for Windows file. The receiving computer then opens the document as a Word file. MIME does not entirely solve the attachment problem. The sending and receiving computers still need compatible software platforms and programs. Some early releases of spreadsheet and word processing programs cannot open newer versions of these programs. The MIME standard can be used for voice, video and graphics programs. This will allow users to send video and audio clips via email.

Two issues with attachments are viruses and network capacity. Residential and business users can have their telephone lines slowed considerably when they receive large multimedia attachments. Moreover, there is a concern that these attachments can contain viruses. Straight text email messages cannot "pollute" PCs with viruses. However, attachments, which are opened from within programs, can harm files if there is a virus included in the attachment.

Privacy on the World Wide Web

According to a survey cited in the June 12, 1997 interactive edition of *The Wall Street Journal*, which was conducted by the Graphic, Visualization, & Usability Center at Atlanta's Georgia Institute of Technology, users ranked loss of privacy as their second biggest concern about using the Internet. (Censorship ranked first.) These people have valid reasons to be concerned about privacy on the World Wide Web.

- *Cookie software:* A site visited by a user can add a small text file to the visitor's browser that identifies the user to the site. The cookie is used for password authentication so people are not required to type in passwords when they visit sites to which they subscribe. It is also used to send personalized research or news leads to users based on their cookie. Sites can track customer behavior based on cookies. The cookie, however, does not give site users' email addresses. Those are provided voluntarily via on-line registration, purchases or contests.

- *On-line registration and contest forms:* Telephone numbers, email addresses, and home and business addresses are collected from on-line forms to various sites. These are frequently put into demographic databases and sold to direct marketers.

- *Powerful databases:* Companies such as Lexis/Nexis and Database Technologies offer on-line services where, for a fee, direct marketers can purchase personal data about consumers.

- *The "wired" universe:* The rise in communicating computers makes databases available to anyone with a modem and a PC. People buy access to lists or they hack into computers containing personal information. As systems acquire more sophisticated security, hackers increase their own break-in capabilities. In 1996, hackers broke into the Department of Defense's computer system. Consumers, however, are aware that computer systems are vulnerable to hackers and are leery of leaving private information on the Internet. .

Privacy Loss on Cable Modems

A computer consultant in a small Massachusetts town, Needham, demonstrated how easy it is to gain access to a neighbor's computer files when cable modems are used for Internet access. The consultant clicked on his neighborhood network icon, and then the "entire network" icon, and finally, on the Microsoft Windows network. His computer screen then displayed 21 domain names, addresses, of neighbors also using cable modems from his cable company. At that point, he roamed freely into files of computers that had been left on. The computers he "roamed" into had not turned off their file-sharing programs.

This episode was written up in the June 8, 1997 edition of *The Boston Globe*. It triggered actions by the cable company, MediaOne, to actively tell customers to turn off their file-sharing programs. MediaOne is also considering other steps such as offering customers filters that would prevent other customers from seeing neighborhood domain names. This incident points out the way cable modems are open to privacy breeches. Cable modems, as pointed out in Chapter 7, work over Ethernet protocols. Ethernet is a "shared" protocol. Everyone shares the same coaxial cable or hybrid coax/fiber path to the cable company's telecommunications equipment. (The cable company's telecommunications equipment is called the headend.)

Moreover, because this telephone line is used for data only and not needed for voice, people tend to leave their computers on all of the time. One person is quoted in *The Boston Globe* article, "The speed is just outstanding, because it makes using the Internet fun again, I think it makes a computer work on the Internet the way most layman think it should. I can now leave the computer connected to the Internet all day long and jump in and out at will." Leaving the computer on all the time makes the computer files vulnerable to attack. It makes precautions such as turning off file-sharing or having a firewall, computer security protection, at the cable company important.

Because Internet software companies see the privacy issue as hurting commerce on the Internet, they are looking at standard software platforms to protect privacy. Microsoft Corporation and Netscape Corporation have agreed to support the same privacy standard. The point of software protection is to offer visitors to sites options on how the information they provide via cookies and on-line registration is used. For example, users could specify that merchants not disclose information to other marketers.

The industry hopes to avoid government regulation by setting standards before regulations are passed. The World Wide Web Consortium, a non-profit Web standards group, is reviewing proposals for privacy standards. In 1996 and 1997, the Federal Trade Commission held hearings on privacy. Moreover, in June of 1996, Representative Edward J. Markey from Massachusetts introduced a bill into Congress that would force businesses to let Web visitors know what information they disclosed to marketers. The bill did not pass. Resolving the privacy issue is an example of the pull between regulation of the Internet by the government, private groups or no one at all. It appears that the federal government is waiting for private, non-profit groups and industry to police itself.

Internet Service Providers and Information Service Providers

Internet service providers sell Internet access, email, Web page hosting and other Internet services such as extranets (see below). Examples of Internet service providers are: UUNET, BBN Planet, PSINet, InternetMCI, Sprint, AT&T, Netcom On-Line Communication Services Inc., Intermedia Communications Inc., cable TV companies and RBOCs. (America Online, Prodigy and Microsoft Network are *information* service providers. They are discussed below.) Internet service providers sell Internet access to residential and commercial customers. Small customers program their modems to dial a telephone number to log into the Internet via an Internet provider. Large customers often have dedicated telephone lines with 24-hour, seven-day-a- week, connections to the Internet.

Internet Service Providers

Many business and non-profit organizations start with "dial-up" connections to the Internet and later change to dedicated access. Quite often, in small and medium-sized companies, staff initially purchase service, individually, from information providers such as CompuServe and AOL. They log on from desktop modems and individual business lines. While the Internet service is a fixed fee, business telephone calls usually cost three or four cents a minute, each user rents a line for $20 per month and they each need their own modem. Eventually, businesses find that they can save money by leasing a high-speed line from an Internet service provider that everyone shares. There is a fixed monthly fee for this line and it connects to the organization's local area network so individual modems are not needed. This is more efficient and provides higher speeds than single telephone lines with modems.

An ISP, Internet service provider, will arrange with a local telephone company for a telephone line, for example a T-1, to the customer. The T-1 is run by the telephone company from the customer to the ISP premises. Other small business or residential customers will dial up a call directly to the Internet provider's modems and data equipment (see Figure 8.1). Large Internet service providers such as UUNET, BBN Corporation, InternetMCI and Sprint have their own high-speed Internet telephone lines. This is called a backbone service.

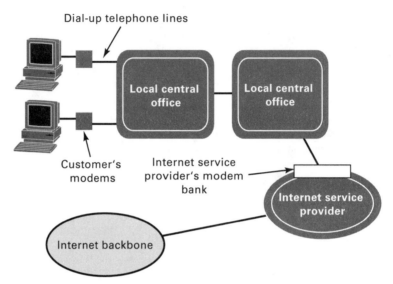

Figure 8.1   A dial-up connection to an Internet service provider.

In addition to large, nationwide Internet service providers, there are scores of local Internet service providers. Smaller ISPs connect their own customers to another Internet provider's backbone Internet network. The customer rents some type of telephone line to access the ISP. The ISP then connects its equipment, via high-speed telephone lines, to the backbone network. UUNET (owned by WorldCom), Sprint and MCI have the largest backbone Internet networks.

Information Service Providers—Content as well as Internet Access

Information service providers, also called commercial on-line services, supply information and entertainment services as well as Internet access. Examples of these information and entertainment services are keyword searches, on-line shopping, games, weather information, profiles of large companies, chat groups, sports data and airline scheduling. AOL has 200 information providers that furnish data for AOL customers. In addition to their own information services, commercial on-line services have links to the Internet. America Online Inc., Prodigy Services Inc., and Microsoft Network are all information service providers.

RBOCs Become Internet Service Providers

RBOCs such as Pacific Telesis (now part of SBC), Ameritech and Bell Atlantic have started offering direct Internet access to consumers. Consider Ameritech's Internet offering. Ameritech announced in January of 1997 that it was selling Internet access for a flat rate of $20 a month through a new subsidiary. Ameritech supplies its Internet customers with customized versions of Netscape's Web browser and CyberPatrol software. The CyberPatrol software is designed to screen out offensive material from children.

Modems that people call into for Internet connections will be installed at Ameritech locations. However, regulatory prohibitions on providing interstate telephone service prevent Ameritech from handling the Internet portion of the calls. The modems will hand traffic off to UUNET Inc.'s high-speed backbone. The backbone consists of long-distance lines for Internet traffic. UUNET also provides backbone telephone lines for AOL and Microsoft Network.

In response to intense competition in the Internet access market, information service providers have lowered their prices to below $30 per month for flat rate, full-time Internet access. Customers are staying connected to their providers for long periods of time because of the absence of connect time fees. This is forcing information service providers to add capacity to their services. On October, 17, 1996, *The Boston Globe* reported that AOL purchased 280,000 new modems to be installed over a period of four years from BBN. The contract between America Online and BBN was for a reported $340 million.

The economic pressure of adding capacity and reliability and the absence of additional connect time fees for their services have forced information providers to find other sources of revenue in addition to subscription fees. Main sources of profitability are shifting to advertising and transactions. For example, America Online now offers pay for play games as well as free games. Customers can play games for $2 per hour.

Internet Addresses

Due to the popularity of the World Wide Web, Internet addresses, as they are currently structured, are running out. Not only are addresses running out, but there are disputes over which organizations should give out Internet addresses. Just as users want particular telephone and toll-free numbers, organizations also perceive Internet addresses as having a marketing value. People want catchy, easy to remember addresses that match their business names. To understand the controversy over addresses, it is necessary to understand the structure of an Internet address.

The Structure of Internet Addresses

A World Wide Web address is known as a URL, uniform resource locator. On Web pages, when users click on highlighted pieces of text and graphics, the browser connects the user to the URL associated with the highlighted text or graphic.

Addressing for these URLs is based on TCP/IP, the underlying suite of Internet protocols. The addresses are called IP addresses. Bits in the address, called octets, represent the network address, e.g., Houston for XYZ Company, and the host computer within the particular network where a home page or document is located. These octets are translated from numeric bits into alphabetical company names and paths to computers within URLs.

Email addresses demonstrate the structure of IP addresses. To the left of the @ sign is the user's name. For example J_Doe@ or J.Doe@. To the right of the @ sign is the domain name. The domain name is the name of the computer at which the email address is located. Universities commonly use lynx with a dot and then an abbreviation of the school's name. To illustrate, Northeastern University's email addresses are username@lynx.neu.edu. Corporations use their name or an abbreviated form of their name for the domain name. To the right of the computer or organizational name is a description of the type of organization using the address. This is a top-level domain. Countries outside of the U.S. use geographic top-level domain names. For instance, one for Bolivia is .bo and that for the United Kingdom is .uk. The U.S. has the following six top level domain names:

- *.com:* commercial businesses.
- *.org:* non-profit organizations.
- *.net:* networks.
- *.edu:* educational institutions.
- *.gov:* governmental bodies.
- *.mil:* the military.

New Internet Address Names

The U.S. is running out of Internet addresses. According to the February 6, 1997 *Wall Street Journal*, 85,000 new Internet addresses are registered each month. Roughly 90% have the top-level domain .com. This is analogous to running out of telephone numbers within an area code. Just as new area codes are created to provide extra telephone numbers, five organizations that oversee the Internet approved a plan, April 8, 1997, to add seven new top-level domains so that additional Internet addresses will be available. Current addresses will stay the same. Organizations applying for new Internet addresses will be assigned the addresses with the new domain names. The new top level domain names are:

- *.arts:* cultural and entertainment entities.
- *.firm:* businesses or firms.
- *.info:* information services.
- *.nom:* private individuals.
- *.rec:* recreation and entertainment.
- *.store:* businesses offering goods to purchase.
- *.web:* entities related to the World Wide Web.

In addition to adding top-level domain names, the governing bodies of the Internet recommended that 28 organizations be appointed to register Internet addresses. Currently, only Network Solutions Inc. registers names.

Electronic Commerce and Advertising on the Web

The nature of the World Wide Web is expanding from a medium for research, collaborative projects and email to one where electronic commerce, advertising and marketing take place. Currently, the dollar volume of products and services sold via the Internet is small compared to that sold via traditional retail and catalogue outlets. Pornography and game sites generate the lion's share of these revenues. However, many industry experts see potential for growth of Web-based sales. Internet users as a whole are more educated and have higher incomes than the general population. Moreover, technical issues such as security and the availability of higher speeds for access to the Internet are expected to lead to increased use of the Internet.

Advertising on the Web—A Source of Revenue

Revenue on the World Wide Web is derived from advertising as well as direct sales of services. Paid advertising is present in the form of banner ads located on the side or top of home pages and publications such as *The Wall Street Journal, Wired Magazine, Playboy* and *The New York Times.* The hope on these small banner ads is that potential consumers will "click" through on the ad and thus visit the advertisers' home pages.

The Internet Advertising Bureau, as reported in the "Technology Briefs" section of the June 16, 1997 *Interactive Edition of The Wall Street Journal,* claimed that first quarter 1997 Internet advertising was $129.5 million, up 333% from $20 million in the first quarter of 1996. Clearly, Internet advertising revenue is rising. However, it is an extremely small portion of the total ad dollars spent in traditional media such as TV, radio, newspapers and magazines.

Certainly, the demographics of Internet users are attractive to advertisers. However, the subject of on-line shopping is a topic of much debate. *Business Week* commissioned a poll by Baruch College and Louis Harris & Associates. The poll was conducted April 11–16, 1997. The following are excerpts from the May 5, 1997 *Business Week*:

> 42% of Internet and Web users have household incomes of more than $50,000 a year....Only 1% of cybercitizens frequently shop online, while 9% do so sometimes....Nearly one-quarter of online users have purchased something either online or on an online service...the Harris survey suggests that senior citizens are most likely to buy online: 42% of those 65 and over have purchased something online...39% of 50 to 64 year olds have purchased something online.

These statistics suggest that while the Internet is not used extensively for on-line purchases, the Internet population is affluent, a large number of Internet users have made on-line purchases and a larger percentage of older people, a growing part of the population, make on-line purchases.

Many sites on the World Wide Web that offer services at no cost to viewers are partially or wholly supported by advertising. An example of these subsidized sites are search engines such as Lycos, Alta Vista and Yahoo. (Search engines are home pages where users type in topics they are interested in learning about. The search engine "searches" the Internet and presents the user with a list of sites pertaining to the topic requested.) There are so many sites on the World Wide Web that search engines are valuable for conducting research and for locating particular homes pages. Therefore, these search engine home pages are visible sites with high usage.

Money Spent to Produce and Maintain Web Sites

Commercial firms that have Web sites spend large sums on the design and maintenance of their home pages on top of hosting and Internet access fees. According to an article in *Wired Magazine*'s December 1996 issue, "Reclaim the Deadzone" by Hunter Madsen, 100,000 companies have Web sites. These sites cost their sponsors $1 billion to $2 billion annually in maintenance and production costs. Mr. Madsen made the following comments about corporate home pages:

> What the zealous captains of commerce have produced so far is a vast and rather eerie necropolis of filigreed mansions, sparsely inhabited. Or perhaps a city of 100,000 childless couples, each pair waiting wistfully by the front door, hoping that you, adored consumer, will knock so they can lovingly welcome you in and take over your life. (Excerpted from "Reclaim the Deadzone" by Hunter Madsen, *Wired*, 4.12 © 1996–1997 Wired Magazine Group, Inc. All rights reserved. Reprinted by permission.)

Web Advertising Pushing Instead of Pulling—Similar to TV

Advertising on the World Wide Web started in 1994. Organizations are debating the best way to advertise on this medium. One school of thought is to make the Web more like television that is a broadcast, passive medium where viewers don't have to click on images to find out more about products and services. The only clicking on TV is to change stations. PointCast Corporation, located in Cupertino, California, uses what is called "push" technology to deliver advertisements to Internet users' PC screens. PointCast offers free screen savers to users. In return, sponsors "push" their ads to users' screens. Screen savers save PC display monitors from burning static images onto monitor screens if the computer is left on for long periods of time without being used. Screen savers generate patterns that appear on PC screens when they are not used for a few minutes.

PointCast promotes itself on its home pages as "the first free network to broadcast up-to-the minute news and information directly to a computer screen via the Internet." To encourage people to use the service, PointCast delivers news on user-defined topics to peoples' computer screens. Advertisements are embedded in the screen saver along with the news. PointCast broadcast services are distributed directly and through Internet service providers in return for a royalty from ISPs.

Another example of push, or broadcast Internet technology, is After Dark. After Dark is a screen saver offered by Sierra On-Line Systems in Bellevue, Washington. It uses "push" technology to deliver ads, stock quotes and news items. It is available free at *The Wall Street Journal* site.

On-line Commerce—What Sells?

On-line commerce is a relatively new source of revenue. Organizations believe that the demographics of Web users point to potential sales on the Internet. Web users have higher incomes and more college education than the general population. Nevertheless, Internet commerce is a risky business. To illustrate the uncertainty of on-line commerce, three large organizations have discontinued their on-line commerce ventures. AT&T started on-line commerce ventures Home Town Network and Industry.Net. Home Town Network was closed in 1994 and a majority portion of Industry.Net was sold to Jim P. Manzi, the founder of Lotus, in June of 1996. Industry.Net, renamed Nets Inc. went bankrupt in 1997. MCI and IBM both started and closed on-line shopping malls.

Adult entertainment, pornography and games are popular services that people pay for on-line. *The Boston Globe* in its December 1, 1996 edition estimated that Internet pornography is a $100 million annual business. According to a *Wall Street Journal* front-page article on May 20, 1997, one adult entertainment site, Amateur Hardcore, was named by PC Meter, a Port Washington, NY company that tracks Web sites, as the 16th most frequently visited Web page. One Web publishing company mentioned in *The Wall Street Journal* article generates revenue of $25,000 a day for live video pornography programming.

Often, game sites are aimed at children. There are good reasons to target children with Internet services. According to figures compiled by New York-based Jupiter Communications Company and printed in the April 14, 1997 *Business Week*:

> When an adult logs on to an online service, he or she spends—on average—an hour. The average kid spends three. Today there are 4.1 million kids surfing the Net. By 2000, there will 19.2 million. Kids, who spent $307 million in 1996 on online services, will spend $1.8 billion by 2002.

Popular sites targeted to children include Microsoft Internet Gaming Zone, Total Entertainment Network (TEN), MPlayer, America Online and Berkeley Systems Bezerk. Internet games are not just for children. One example of an adult on-line game is electronic pinball developed by PlayNet Technologies in New York City. PlayNet pinball machines are used in restaurants and bars. Customers pay to play pinball against other on-line users. .

On-line Internet commerce is still in its infancy. Organizations are using the Internet to add value to products that are sold via retail channels. Businesses such as Amazon.com,

Up-selling Classified Ads for On-line Newspapers

Community Newspaper Company, a newspaper chain in eastern Massachusetts that owns 120 local newspapers, has a presence on the World Wide Web. It is the Internet edition of their newspapers, *Town Online*. When customers call Community Newspaper Company's classified advertising group to place an ad, they are asked if they would like the ad to appear in both the on-line and traditional paper editions for an additional fee. The on-line classifieds offer surfers the opportunity to directly access five categories of classified ads: real estate, vehicles, employment, computers and merchandise. *Town Online* hopes to attract Internet users to their site with their emphasis on local content. They offer local news, calendar information, restaurant reviews, education listings and community resource facts. Readers have an opportunity to post comments about a variety of issues in the "Bulletin Board" section.

founded in 1995 in Seattle, Washington, use the Internet as their only channel for sales. Amazon adds value to book purchasing by its many on-line book reviews and interviews with authors. They offer discounts, a large stock of inventory, 2.5 million titles, and next day delivery. On-line brokerage firms offer value in the low fees they charge for stock trades; fees that are below those of discount brokers.

Other industries offer on-line services that are adjuncts to their standard services. Newspaper and journal publishers have on-line editions that are supplements to their hard copy, paper versions. One advantage of on-line versions is that columns can be sorted by topic. For example, readers can be presented with all of the articles about one topic, for example, technology, without thumbing through a whole newspaper. On-line editions have the capability to be sorted by areas of interest. In addition, research and links to related sites are available with on-line publications. They are not "stand-alone" publications. Readers can follow "threads" on topics. For example, an article in *The Wall Street Journal Online Edition* on privacy on the Internet had links to groups that monitor privacy on the Internet. They also offered readers the opportunity to participate in on-line discussions on privacy on the Internet.

Web Congestion—World Wide Wait

The term World Wide Wait was printed in the March 24, 1997 issue of *Network World*. A major frustration of using the World Wide Web is waiting for pages to be downloaded from sites. Making the Internet easy to use and adding graphics has created congestion and speed problems. Browsers make surfing the Internet easy, but they contain a large number of bytes. Browsers, graphics-rich games, speech and the video found on some sites contain large numbers of computer bits. This makes viewing pages slow and tedious and hinders acceptance of the Web.

Delay is a problem that must to be addressed for both the adult entertainment and game segment of Internet commerce and for Web usage in general. The speed of downloading video and high-bandwidth graphics from the Internet is uneven. Delays are present in both the Internet portion of the traffic and in lines linking customers with their telephone companies. Traffic on the Internet travels in little packages called packets. Packets can be delayed because of congestion, or because of the routing of packets between Internet operator networks. In May of 1997, nine of the major Internet operators voluntarily formed a consortium, IOPS, Internet Operators' Providers Services, to establish procedures to make routing traffic between their networks more reliable. It is not unusual for traffic to flow through three or more networks. If one of these networks routes packets to a congested or out-of-service link, the packet either arrives slowly or must be retransmitted.

In addition to forming the consortium, network operators are spending large amounts of money to upgrade their telephone equipment and lines. GTE bought BBN for its Internet network. AT&T, MCI and UUNET have all announced major capital expenditures to upgrade their telecommunications facilities. For example, in February of 1997, WorldCom Incorporated announced they were investing $300 in their subsidiary UUNET's Internet network. The investment is to upgrade the speed of their network from T-3, 45 million bps, to ATM, 622 million bps.

Even when packets travel quickly and reliably through the Internet, local telephone lines are slow. Most residential consumers have 28.8 thousand bit per second or slower analog modems. The performance of these analog lines from end-users' homes to their telcos are uneven at best. In some cases, consumers are willing to upgrade, but digital services such as ISDN are not available. Often, higher speed local lines are available, but users do not want to pay for them. Hopefully, increased competition in local calling will increase availability of high-speed services and lower their prices.

Having users convert to higher speed telephone services is not just a technical and economic issue. Consumer education is another factor in the decision to upgrade to higher speed modems and digital telephone lines. Installing a new modem or ISDN service can mean changing communications software and making programming changes. Not everyone has the computer literacy to make these changes. There are many factors that impact acceptance of the Internet as a place to shop and experience entertainment. Certainly, having Web pages that download at an acceptable speed is a major factor.

Intranets—Impact of Web Technology on Internal Operations

An intranet is the use of World Wide Web technologies for internal operations. Employees use browsers on their PCs for applications such as collaboration on projects and looking up employee extension numbers. An intranet is the use of Web-type browsers to provide employees access to internal information. However, unlike the Internet, outside users cannot access intranet applications. Security is built into these applications such that only authorized users have access to the internal databases and documents. Even though external devices do not have access to internal intranets, internal users can get to the Internet.

According to *Business Week*'s February 2, 1997 article "Netspeed at Netscape", the market for intranet software is expected to be $10 billion by the year 2000. Industry analysts generally agree that the market for intranets is growing. The following are ways that organizations use intranets:

- Project monitoring.
- Project updates.
- Publication of regulatory manuals.
- Internal job postings.
- Internal telephone books.
- Distribution of custom-made software applications.
- Distribution of training schedules.
- Vehicle for signing up for training classes.
- Conference room scheduling.

Hard copy internal telephone books are an enormous frustration at Fortune 500 businesses. As soon as they are published, they are out of date because employees frequently move and change locations. This is a perfect application for an intranet. Posting telephone numbers on a browser puts changes at people's fingertips as soon as they are logged. Moreover, staff can search by keyword for functions, location addresses, mail stops and departments.

The advantage of the World Wide Web browser technology is that it can work with existing databases. Servers, computers, with the Web protocol, HTTP, contain text or keyword indexing to work with in-place databases. On-site databases do not have to be changed to add intranet services.

Finally, the growth of intranets adds more traffic to already congested commercial, government and non-profits' internal networks. Browsers are "bandwidth hogs". They have color, sound and graphics capabilities. They add traffic to local area networks, campus connections between LANs and connections between LANs across the U.S. and worldwide.

Extranets—Using Internet Technology with Customers, Partners and Vendors

Extranets extend the reach of intranets from internal-only communications to sharing documents and information for business-to-business transactions. Typically, the other businesses with which information is shared are suppliers, vendors and trading partners. Placing orders is one opportunity for extranet services. Existing customers use passwords to log into suppliers' databases to check on availability and rates for products. They can place and check the status of orders from the same Web browsers they use to check on pricing and availability. Often, these orders are transmitted directly to mainframe computers without human intervention. The absence of human intervention eliminates the possibility of human error in input.

Benefits of Web technology for extranet applications are:

- *Decrease labor costs:* Customers place their own orders.
- *Save on paper:* Customers print their own orders.
- *Shorten ordering cycles:* Customers can access pricing, rates and delivery dates without waiting for salespeople or customer service representatives to let them know the information.

Another extranet application is trouble ticket reporting. Large organizations such as Lucent Technology give customers the ability to report repair problems via the Internet. After a customer logs in with his or her identity such as his/her customer ID number or location, he/she fills out a form describing the repair problem. The trouble ticket assigns him/her a trouble ticket number for tracking purposes. Both the customer and maintenance organization now have an electronic timestamp of when the trouble was reported. The trouble ticket is automatically sent to the dispatch staff to be resolved.

Doing business using Internet technology assumes a computer-literate business partner. If that is the case, Web browsers are a universal-type interface. However, in some cases, customers are not computer-literate. One such example involves Mobil Corporation's effort to have its lubricant dealers order supplies over the Web. A story in the November 1996 *Wall Street Journal's, "Technology"* section pointed out that Mobil Corporation went to great lengths to train its dealers. The article quotes Mobil's Russ Rieling, head of the Internet/intranet strategy group:

> We had to train them and hand-hold them and give them a lot of care,…But we haven't met a lot of resistance.

Mobil Corporation went a step further than hand-holding and training. It offered to sell its dealers PCs preconfigured with Web browsers. Configuring a PC with the protocols and software required to access the Internet is more difficult than using a computer.

A concern with extranets is security. Extranet applications give outside organizations access to portions of concerns' databases. Three security issues addressed on extranets are:

1. *Authentication:* Authentication assures the receiver that the sender is who he/she claims to be and not a hacker.
2. *Integrity:* Integrity checks assure the sender that no third party has inserted third-party data such as viruses that damage corporate data. Integrity checks that the data is what is claims to be and not something that can harm computer files.
3. *Encryption:* Encryption scrambles the data sent so that no one except the intended recipient can "read" the data.

Security on the World Wide Web— Establishing Trust that the Internet Is a Secure Place for Transactions

Doing business on the Internet is a new way to interact in business transactions. The face-to-face and telephone-to-agent interactions that provide assurance to customers are absent on the Internet. People who shop in stores have a personal encounter with clerks. They assume the salesperson who takes their credit card is trustworthy because they see him or her. They know that person is authorized to take their money, debit card or credit card. People who shop on the Internet need to find a way to establish trust with the organizations with which they do business. Establishing security on the Internet will help increase its utilization for electronic commerce.

Tools that are used for security on the Internet are used in private networks as well as the Internet. These tools are public key encryptions. Public key encryption mathematically encodes documents so that they cannot be read by anyone except the intended recipient. Encryption scrambles documents using mathematical algorithms so that only the intended recipient can decrypt and read the document with its matching private key.

To authenticate the user, a digital signature is sent along with the encrypted document. The digital signature verifies that the person is who he or she claims to be by sending a digital summary of the data sent. The receiving end receives the digital signature and makes a mathematical summary of it. If the receiving end's summary exactly matches the sending end's summary, the identity of the sender is verified.

Security tools, known as firewalls, are installed in corporate and Internet service provider computers in front of corporate databases. Firewalls screen transmissions before allowing them to reach corporate computers. The firewall verifies the integrity of the data and the sender. Many organizations outsource their Internet applications to Internet service providers. When they do, they want assurance from the Internet providers that the outsourced applications are secure. Firewalls are installed extensively at Internet service providers as well as end-user sites.

Making the Internet a Trusted Place to Do Business

Chrysalis-ITS in Ottawa, Canada makes cryptographic products for secure electronic transactions on Internets and intranets. According to their President, Steve Baker, "Chrysalis-ITS helps improve the security and performance of the Internet. We are one of many companies doing things to make the Internet more useful." Steve feels that banks and financial institutions will drive acceptance of the Internet as a secure place to transact business. He thinks that customers trust banks and are not afraid to give them money. He believes that people also trust airlines, although not as much as banks, as places to spend money. Currently, Wells Fargo in the U.S. and Royal Bank in Canada are conducting financial transactions with customers over the Internet.

An additional form of encryption security is tunneling. Tunneling encapsulates encrypted packets within other protocols for added security through virtual private networks such as the Internet. Tunneling separates and keeps private transmissions from multiple customers. Tunneling allows new protocols to be packaged and transmitted within older protocols. They are "unwrapped" when they are received by remote firewalls at extranet and intranet sites.

Conclusion—Reliability and Capacity

The biggest single factor impeding acceptance of the Internet is the slowness in downloading and moving around World Wide Web documents. The bottlenecks exist at the local loop from customer premises to Internet service providers. Two problems exist in the local loop.

1. Most telco services are analog POTs, plain old telephone service (analog).
2. Connections from end-users to Internet service providers are circuit switched services designed for short voice calls.

95% of residential customers still have analog modems with analog lines. ISDN is available in many locations, however, users don't understand how to install it or what type of equipment and programming is needed to implement ISDN. Moreover, ISDN, while faster than analog POTs service, is slow for streaming video that will become more prevalent in sites with games and entertainment services.

In addition to a majority of the residential telephone lines being analog, they are connected to central office switches designed for voice traffic. With voice traffic, calls are short, about three minutes in length, and only one out of eight users is generally on his/her telephone at any given time. For this reason, central office switches are designed such that between one out of four or one out of eight users can be on the telephone at any given time. As more people use their telephone lines for data and stay on calls longer, users may start to encounter network busy conditions.

POTs and ISDN telephone connections are circuit switched connections; when one person uses his/her telephone, the path between the telephone and local telephone company cannot be used by anyone else. This is true even during pauses in conversation. These circuit switched connections are not an efficient use of telephone services. They do not allow the same path to be shared by multiple users.

In contrast, most data networks, including the Internet, work on some type of "sharing" scheme. The Internet protocol, TCP/IP, breaks up all transmissions into packets. Packets from many different data devices share the same telephone path. In essence, the local telephone switches, which were designed for voice traffic, are now being used, inefficiently, for data. As bandwidth requirements increase, efficient use of facilities will become more important.

The need to upgrade facilities is part of an issue involving regulatory controversy. Bell telephone companies and ACTA, America's Carriers Telecommunications Association, have urged the FCC to allow local telcos to charge Internet service providers surcharges, in the form of per minute access fees, for carrying this extra Internet traffic. The local telephone companies claim they need extra revenue to upgrade their facilities. The interexchange carriers, represented by ACTA, claim that since the Internet is used to carry voice and data telecommunications traffic, Internet service providers should pay the same access fees as carriers. They should be treated as though they are carriers.

The carriers, local telcos and Internet service providers have issued documents making a case for their side. The Bells have produced studies that indicate central offices are running out of capacity. The Internet service providers have produced studies stating that these claims are exaggerated. Meanwhile, The Network Reliability and Interoperability Council, an industry group that tracks reliability of the public network, stated that Internet traffic has, up to the first quarter of 1997, not caused any outages in the public network. The FCC ruled that Internet service providers are not required to pay access fees to local telcos.

XDSL technology, discussed in Chapter 7, is a leading contender for Internet access that provides both high-speed telephone connections and the ability for telephone companies to separate voice and data transmissions *before* they are passed through a central office switch. With XDSL, data calls will not congest local switches. To become a viable alternative, XDSL modems must come down in price and local lines from users to telephone offices must meet technical requirements. These requirements are met in about 80% of the local telephone lines.

It's unclear what the role of competition and The Telecommunications Act of 1996 will play in spurring the availability of high-speed services in local telephone services. The Act mandates universal availability of access to advanced data networks such as the Internet. It also mandates the connection of new local exchange carriers to elements of the Bell telcos. Just as divestiture in 1984 stimulated competition and thus the availability of advanced services by long distance companies, perhaps competition will help make high-speed services and capacity available on the local level.

Reliability, in addition to speed and capacity, is a concern as Internet usage grows. The FCC oversees reliability of the public switched telephone network. Any outage affecting more than 3,000 users must be reported by telcos. Despite the federal government's funding of the original Internet, the FCC has never regulated the Internet. Instead, the industry has created voluntary groups to police themselves. A case in point is IOPS.ORG, Internet Operators' Providers Services, founded by Internet network operators to agree on ways to route traffic to each other. IOPS will also look at trouble ticket reporting between networks on the Internet backbone. Their hope is to police themselves so that government oversight will not be required.

Finally, upgraded Internet protocols are being reviewed by Internet governing bodies that will enhance World Wide Web functions in addressing, electronic mail, sending video through the Internet and sending voice traffic through the Internet. Currently, the Internet is a work-in-progress with new applications such as voice, video and graphics. A robust Internet will help ensure its success.

CHAPTER 9

Wireless Services

Wireless services are being developed to serve many diverse telecommunications markets. These services include paging, analog cellular, digital cellular, personal communications services, PCS, low earth orbiting satellites and specialized mobile radio. Some of these technologies, such as paging, specialized mobile radio for dispatch and delivery functions and cellular telephones, have been available for more than ten years. Others, such as low earth orbiting satellites, are new or still in development. Companies from all over the world are putting money into research and development efforts for new cellular and satellite technologies. The goal of this chapter is to sort out the various wireless options, which are described in Table 9.1.

Existing analog cellular services in the U.S. provided by local telcos and their competitors were developed in the 1970s by AT&T. They started to be implemented in the late 1980s. They were implemented in a standard format and all telephones work with all analog cellular offerings. The most important function of analog cellular technology was to add capacity to existing non-cellular mobile telephone service. It made car telephones affordable to corporations initially, and later as prices decreased, to residential consumers.

However, analog cellular services became so popular that there was a concern that capacity would be depleted, particularly in metropolitan areas. As a result, digital cellular multiplexing technologies were developed to add capacity. Digital cellular also offers customers added features such as caller ID, call forwarding and three-way calling, as well as handsets with paging and short messaging services integrated into the handset via a liquid crystal display. Digital multiplexing techniques were planned to be used on existing cellular airwave capacity, as well as on new air space to be distributed by the FCC.

All digital cellular services use multiplexing so that diverse devices can share the same wireless channels. Digital cellular services are being implemented differently in the U.S. than in Europe. In 1987, the European Union chose a standard called GSM for delivering new digital wireless telephony. In the U.S., the TIA, Telecommunications Industry

Table 9.1   Wireless Services

Service	Frequencies	Features	Comments
Original analog cellular telephony	824 to 893 MHz	Provides basic calling and voice mail. Added capacity when implemented in the late 1980s .	Each area has two providers, the local telco and a competitor. All telephones can be used on all services.
Digital cellular telephony	824 to 893 MHz	Provides advanced features such as caller ID and telephones with built-in pagers. Gives carriers more capacity on existing channels.	Most Bell companies and AirTouch use CDMA. SBC, BellSouth, AT&T Wireless and Bell Mobility use TDMA.
GSM, global system for mobile communications	890 to 960 MHz	Users in these countries can use their wireless telephones in all of these countries.	Standard used in Europe, the Far East, Israel, New Zealand and Australia.
PCS, personal communications service	1.8 to 1.9 GHz	Prices to consumers are lower than cellular. Has same features as digital cellular. Uses more closely spaced, smaller antennas than cellular.	Sprint Cellular, AirTouch, PrimeCo and Alltel use CDMA. AT&T Wireless uses TDMA, which is incompatible with CDMA.
LEOs, low earth orbiting satellites 500 to 1000 miles above the earth	1.6 GHz	A way to supply wireless telephone, Internet access, video and data communications to remote locations.	Systems under development by Motorola, Craig McCaw's Teledesic Corp., Loral and others.
SMR, specialized mobile radio	800 to 900 MHz	Offerings include packet data from RAM and ARDIS and Nextel's wireless digital telephone service.	Originally used for analog, two-way dispatch services.

Association, settled on a similar standard using time division multiplexing. However, shortly thereafter, many of the Bell telephone companies decided to use a newer method of multiplexing, code division multiple access, which promised greater capacity than time division multiplexing. Thus, the U.S. started down the road with two different standards, both different than Europe's. Interestingly, there were seven different, incompatible analog types of cellular service in Europe before digital GSM was installed. Incompatibility is a problem because customers with incompatible digital wireless telephones cannot use their telephones when they travel to places with incompatible cellular service.

There are two types of digital wireless service in the U.S. They are different in that they use a different frequency or portion of the airwaves to transmit signals. One is called PCS, personal communications service, and the other is known as digital cellular.

- Both can utilize either code division or time division multiplexing.
- Both provide roughly the same telephone features.
- Both are more difficult to eavesdrop on because the multiplexing scrambles the voice signals.
- PCS is lower in cost.
- PCS towers are smaller and must be spaced closer together.

Another wireless service that is well-established, in addition to analog cellular, is paging. Sales of paging services are booming. More and more consumers and commercial users alike expect to be reachable at all times. Paging rentals have expanded from primarily professionals to consumers. Moreover, their functionality has been expanded. In particular, voice mail and two-way messaging are being made available. Also, unlike new digital telephone services which are concentrated in metropolitan areas, pagers that can be reached from many remote areas of the U.S. are on the market.

Accessibility from remote regions is one of the primary motivations for the development of low earth orbiting satellites, LEOs. Companies such as Motorola, Loral, Hughes, TRW and Teledesic Corporation have poured billions of dollars into the purchase of airwave licenses and the development of technology to be used by satellites which will orbit the earth at lower altitudes than current satellites. The goal is to provide telephone, data communications, broadcast-quality video and Internet access to remote corners of the world. It is still unknown if and when these efforts will be profitable.

Another wireless technology discussed in this chapter is specialized mobile radio, SMR. SMR was used originally for dispatch services in businesses such as contracting. The airwaves on which these channels were used were later deployed to transmit data for transportation, field maintenance and delivery organizations. These are the ARDIS and RAM networks backed by Motorola, IBM and BellSouth. Finally, Nextel changed its analog data network services from all data communications to a voice and data network. They upgraded analog services to digital and now sell digital telephone services with the functionality of digital PCS cellular service. Table 9.1 provides a comparison of the various wireless services available.

Historical Background of Mobile and Cellular Services

Prior to the availability of analog cellular car telephones in 1984, users who wanted to place telephone calls from their cars used mobile telephones. The first mobile telephone system was started in 1946 in St. Louis, Missouri. Costs for car telephones were high, in the $2,000–$2500 range, and capacity was limited. Each city had one transmitter and receiver for the entire area run by the local telephone company. Thus, the entire area covered by the one transmitter shared the same channels. This meant that only a limited number of simultaneous calls could be placed on each city's mobile system. This equated to a range of 25 to 35 concurrent calls. In addition to limited capacity, the quality of service was spotty with considerable static and breaking up of calls.

More widespread than mobile telephone service prior to the mid-1980s was mobile radio service. Mobile radio is a "closed" service. Mobile radio operators can call each other. They cannot, however, make calls to the public network or to users on other radio services outside of their own system. Calls are made to other people on the same mobile radio system. For example, users on one taxicab service's system cannot call users on another cab's system. Police departments were early pioneers of car radios. The Detroit police department used mobile radio in 1921. In the 1930s, mobile radio use spread to other public safety agencies such as fire departments. Mobile radio systems are now used for aviation, trucking, taxis and marine applications.

Mobile radio is half duplex, calls are two-way but only one way at a time. For example, when one person is done speaking, he/she uses a convention such as "over and out" to let the other person know he/she is through talking. People using mobile radio "push" a button to talk. A prime example of a mobile radio system, not linked to the public telephone network, is a system used by a taxicab company.

Spectrum Allocation

Aviation, marine, trucking and emergency public agencies all use different portions of the spectrum, or airwaves, to communicate over wireless radio service. The term "spectrum" refers to a range of radio frequencies or the portion of the radio waves used to make telephone calls or transmit data. A frequency is the number of times each second that each radio wave completes a cycle. As mentioned in Chapter 1, each cycle looks like a resting letter S. A cycle is complete when a radio wave passes through the highest and lowest portions of the wave. The term "hertz" refers to one cycle of a radio wave.

Spectrum is broken up into bands and assigned by the FCC for particular purposes. For example, residential cordless phones are assigned to the 46 and 49 million cycle (hertz) per second, MHz, bands. Citizens radio is assigned to the 27, 462 and 467 MHz, or million hertz frequency, bands. Consider the very high frequency band which TV channels 2 through 12 and mobile radio services such as police and fire dispatch use. This band is in the 30 MHz to 300 MHz, or million hertz, range of frequencies. Thirty MHz means that each wave has 30 million cycles or hertz per second. It completes 30 million cycles; in other words, it has 30 million resting Ss in one second.

The higher the frequency, the smaller the radio wave. For example, a 3000-hertz wave is longer than a 3,000,000-hertz wave. Small waves are more susceptible to rain and weather conditions. A rain droplet can destroy a smaller wave more easily than a larger one. The raindrop is bigger in relationship to a small wave than to a large wave. For this reason, high-frequency microwave systems are more susceptible to weather conditions than lower frequency systems. Higher frequency services such as personal communications services, PCS, also cannot be transmitted as far as lower frequencies. Therefore, PCS towers and antennas must be closer together than lower frequency, traditional, cellular services covering the same area.

Because there is a limited amount of radio frequency, or spectrum, and to eliminate interference, the FCC allocates frequencies. If frequencies were not allocated to specific companies for specific uses, transmissions would overlap or interfere with each other. For example, if two telephone calls took place in the same airspace over the same frequency, the callers would be able to hear each others' conversations.

Because of the inherent limit on spectrum, a goal of new wireless technologies is to do more with less. In other words, newer technologies find ways to share spectrum with an increasing number of users without degrading the quality of service. When analog cellular was introduced in the 1980s, spectrum was "stretched" by reusing the same frequencies over and over. This was accomplished by setting up hexagonal-shaped cells. Cells that adjoin each other use different frequencies. However, cells that do not abut use the same frequency already used in a cell which is not adjoining.

The next improvement in spectrum sharing from analog cellular is digital cellular and PCS. Digital cellular packs more capacity on channels by multiplexing. The two multiplexing technologies used in digital cellular are time division multiplexing and code division multiplexing.

Cellular Telephone Service—Advanced Mobile Phone Services

Cellular telephone service, or advanced mobile phone service, is the analog cellular telephone service available in urban and rural areas of the U.S. It is the next stage in the development of car telephone service from mobile telephone service. Mobile telephone service has severe quality of service and capacity limitations. A very limited number of conversations, often only 25 to 35, can take place at the same time in each metropolitan region. Cellular telephone service has more capacity because it reuses frequencies in hexagonal-shaped cells depicted in Figure 9.1. Each cell can have up to 57 conversations per cell. If carriers need more capacity, they split cells into smaller sizes and reuse their frequencies. More cells equal more capacity.

The concept of cellular telephone service originated in 1947 at AT&T's Bell Laboratories. The first cellular telephone systems were trialed in Chicago and in Baltimore, Maryland in late 1983. Meanwhile, the FCC set aside radio spectrum for cellular service at 825 MHz to 890 MHz. Each of the 306 Metropolitan Statistical Areas and 428 Rural Service Areas were to have two cellular service providers: one would be the local wireline

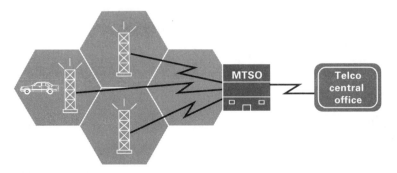

Figure 9.1   The mobile telephone switching office, MTSO, connects the cellular network to the public switched telephone network.

telephone company, the B block of frequencies; and the other, a non-telephone company, was assigned the A block of frequencies. By having two providers in each area, the FCC hoped to foster competition. Indeed, the cost for cellular handsets quickly dropped and their size decreased. The telephones are often provided at no cost or at a minimal fee to lure new customers.

Other than the 30 largest metropolitan areas, the non-wireline frequencies were given out by a lottery held by the FCC for qualified vendors. By 1987, 200 cellular systems were in place in 127 cities. Initially, cellular service was purchased for salespeople and business executives who justified the high cost of telephone calls by their ability to use their time more effectively by reaching customers while in their automobiles. The cost of each call was roughly equivalent to the cost for an operator-assisted call on a traditional landline telephone. Corporations routinely spent $200 per month on salespeople who were frequent users of car phones.

The Cellular Market

Analog cellular telephone service was slow to become popular in the U.S., as demonstrated by the statistics in Table 9.2. The large increases in the number of subscribers began in 1990. The statistics illustrate growth in the number of subscribers, but decreases in average local telephone bills from 1988 to December of 1996. The year 1996 includes PCS services, most of which were only in service for part of the year.

While the number of subscribers has grown 88% from 5 million in 1990 to 44 million in 1996, annual revenues have only grown 81% without factoring in inflation. Some analysts blame the fact that average monthly local cellular phone bills are decreasing on the high cellular per minute costs. *The Boston Globe*'s July 12, 1996 article quotes Business Research Group in Newton, Massachusetts on the low average usage of cellular users: "The primary inhibitor is service costs." The same article states that, according to Yankee Group, the average number of minutes per month is 90–100. In Western Europe, it is 150–200. The penetration of cellular telephones is also lower in the U.S. than in Europe. Analysts blame the lower number of cellular telephones on the higher U.S. fees for cellular service.

Table 9.2   Cellular Telephone Usage

Date	Estimated Subscribers	Annual Revenues (000s $)	Average Local Monthly Bill
December 1988	2,069,441	1,959,548	$98.02
December 1989	3,508,944	3,340,595	$83.94
December 1990	5,283,055	4,548,882	$80.90
December 1991	7,557,148	5,708,522	$72.74
December 1992	11,032,753	7,822,726	$68.68
December 1993	16,009,461	10,892,165	$61.48
December 1994	24,134,421	14,229,920	$56.21
December 1995	33,785,661	19,071,966	$51.00
December 1996	44,042,992	23,634,971	$47.70

Source: The CTIA Semi-annual Data Survey. Used with permission of CTIA.

One thought on the reason for lower minutes of usage per month is that initial users were business customers, in particular, salespeople and executives. These customers have high numbers of calls. As consumers enter the market, they are more cost-conscious and purchase the telephones more for safety reasons than the need to make daily calls. Because competition is increasing with the entrance of new digital personal communications services, PCS, per minute prices may drop. New non-PCS competitor Nextel dropped its roaming fees in January of 1997. Roaming services allow customers to use their cellular telephones in other than their home regions.

Although analog cellular service has been available since 1984, service is not evenly good. Many customers complain about choppy calls and locations where their telephones do not work. Tower spacing, capacity and interference in the form of tunnels and low-lying areas are still problems. Cellular providers are faced with the challenge of finely tuning their transmitters to eliminate dead zones and holes in coverage.

Limitations of Switched Circuit Data Transmission on Analog Cellular

Like the public switched telephone network, the analog cellular network, designed in the 1970s by AT&T, was created for voice services. According to Motorola, less than 50% of people who have credit card sized modems for portable computers use them with wireless analog cellular services. The major impediment to using modems with analog cellular services is the change in signals and the errors introduced during the handoffs

Bell Companies' Cellular Out-of-region Expansion

Southwestern Bell Mobile Systems Inc. provides cellular service in the SBC region of Arkansas, Kansas, Oklahoma, Missouri and Texas. They also sell cellular service to 110 million people out-of-region under their CellularOne name. In 1987, SBC purchased the CellularOne name and cellular operations in Washington/Baltimore, Chicago and Boston. In 1989, it joined forces with McCaw Cellular (bought by AT&T Wireless in 1993), and in 1992, with Vanguard Cellular Systems, Inc. CellularOne is run as a franchise, with members in the above regions plus upstate New York. It has 65 member companies.

SBC states that its goal is to become a full-service provider of wireless, paging, local, long distance and Internet access in every area in which it has cellular service. On January 8, 1997, it announced that it was offering local telephone service in Rochester, NY. It also offers long distance in the following cities in which it sells cellular services: Chicago, Boston, Washington, DC. and Rochester, NY. Having a presence as a cellular vendor may be a leg up for sales of out-of-region telecommunications services.

Bell Atlantic gained market share outside of its region by buying NYNEX Mobile Communications outright, which covers the states of New York, Massachusetts, Rhode Island, Vermont and New Hampshire. BellSouth, on the other hand, expanded out of its region by purchasing independent carriers in parts of Wisconsin, Indiana, Illinois, Virginia, Hawaii, and California.

between base stations and when the signal is transmitted between the mobile portion of the network and the landline-based portion of the network.

Because of the delays and constant retransmissions due to errors, modems, when used over cellular services, rarely transmit at their top speed of 14.4 Kbps. The most common speed tends to be 9600 bps, far lower than speeds generally achieved over wire lines.

When analog cellular is used for facsimile transmission, lines of text are frequently dropped and black lines are sent instead of the images transmitted. This is due to the half duplex nature of facsimile. With half duplex services, transmissions are only in one direction at a time. While the fax is being sent, no error detection messages can be sent from the receiving fax machine to the sending unit. Thus, the facsimile modem keeps sending the document because it does not know errors have occurred. The errors are in the form of skipped and black lines.

CDPD—Cellular Digital Packet Data over Cellular

Cellular digital packet data, CDPD, was first offered in 1995. It was developed by IBM as a way to transmit wireless data over spare capacity in cellular providers' analog networks. It was envisioned as a way to inexpensively transmit short, bursty-type messages such as electronic mail, credit card verification, alarm monitoring and dispatch communications. Because it uses spare capacity in cellular networks, it was planned as a low-cost solution for people who needed to transmit short messages from multiple locations. Pricing as of June of 1996 for services from AT&T Wireless, Bell Atlantic NYNEX

Mobile and GTE Mobilnet ranged in the $90 to $145 per month per user for close to full-time usage.

As of June of 1996, the service was available in 44 U.S. markets. Users can roam between Ameritech, GTE Mobilnet and AT&T Wireless service areas. CDPD, cellular digital packet data was highly touted in the media prior to its introduction. A *Wall Street Journal* article on February 11, 1994 quoted industry analysts as saying, "CDPD ends the argument,…it has the best chance to become a de facto standard in the industry." Contrary to these expectations, CDPD, according to Frost & Sullivan statistics, captured only 1% of the mobile data market in 1996.

Possible reasons for slow acceptance of the service include initial high prices for equipment, customer distrust of new service and lack of universal availability. Finally, users know that digital personal communications service, PCS, networks will offer higher speeds for data communications. PCS networks are newer and were designed taking data transmissions requirements into consideration.

Digital Cellular Telephony—D-AMPS

During the 1990s, analog cellular service gained in popularity. Cellular carriers were faced with the need to add capacity. Initially they added capacity by breaking cells up into smaller sizes. Adding cells adds capacity as each new cell handles an additional number of simultaneous calls. However, adding cells is not without problems. Smaller cells lead to more dropped calls and dead areas where calls cannot be made because of problems with overlapping into adjacent cells. Caller dissatisfaction with quality of service is a big problem for cellular providers. It leads to customers discontinuing service or changing carriers.

The need to add capacity is a major impetus in existing carriers' moves to upgrade their analog cellular service to digital. Another motivation is the desire to offer customers enhanced features. Digital cellular handsets support the same features available in PCS services: caller ID, call waiting, alphanumeric paging, a longer battery life and other calling services such as return call. Carriers hope to garner extra revenue from the sale of these enhanced features.

Privacy as well as capacity is improved in digital cellular. Analog cellular signals are easily listened in on by snoopers with scanners. Cases in point are the famous Princess Diana romantic conversation and Newt Gingrich's political discussion that were reported in the media. Listening in on digital transmissions is more difficult because the digital bits are scrambled when they are multiplexed using time division multiple access, TDMA, and code division multiple access, CDMA, schemes. However, even though it is more difficult to listen in on digital conversations, privacy is not guaranteed. Sophisticated digital scanners can be used to listen in on cellular digital calls. Nevertheless, digital scanners are more expensive than analog scanners. They are also not as readily available. Many vendors of digital scanners claim they only sell them to law enforcement agencies.

Encoders in handsets decode the digital bits representing caller ID numbers and paging messages into alphanumeric characters that can be displayed on handsets' liquid crystal displays. Handsets for digital cellular have both an analog and digital mode; they are dual mode. When callers roam into areas without digital cellular, their calls are automati-

Use with Caution

A study of automobile collisions in Toronto published in the February 13, 1997 issue of *The New England Journal of Medicine* by doctors D.A. Redelmeir and R.J. Tibshirani found that, "The risk of collision when using a cellular phone was four times greater than when a cellular phone was not being used." The authors observed no safety advantage to hands-free telephones.

In an editorial on the subject published in the same issue, Doctors M. MacLure and M.A. Mittleman argue that because of the small size of the study, 699 drivers, "The question of whether telephones that allow the hands to remain free are safer than hand-held telephones remains unanswered."

Laws against using hand-held telephones while driving exist in Brazil, Israel, Australia and Switzerland. MacLure and Mittleman stress the need for more research. They urge the industry to include warnings and advice with their products and bills. Redelmeir and Tibshirani suggest that because drivers who use a cellular telephone are at increased risk for a motor vehicle collision, they should consider these strategies:

- Refraining from making or receiving unnecessary calls.
- Keeping calls brief.
- Interrupting telephone conversations if necessary.
- Minimizing other distractions.

cally put into the analog mode. They do, however, lose their caller ID option and other advanced features associated with digital cellular.

Traditional analog cellular suppliers such as Bell Atlantic NYNEX Mobile and CellularOne offer digital cellular service over the same frequencies as their analog cellular services. Depending on the digital technology, digital cellular has from three to ten times greater capacity than analog service. Cellular carriers offer digital cellular on the same frequencies as their analog service. They set aside channels for digital service which use either time division multiplexing or code division multiplexing. (See below for an explanation of these multiplexing schemes.) With TDMA, existing towers and mobile central offices can be used with new digital transmitters and receivers at the cell sites to carry the digital signals. Carriers designate a portion of their total channels to transport calls in digital format. According to CellularOne in Boston, approximately 20% of its channels are allocated to digital service. Code division multiplexing, however, requires extensive upgrades to the cellular infrastructure.

Alphanumeric paging with telephone number and short message display capability and enhanced features such as caller ID are made possible by Signaling System 7. As explained in Chapter 5, SS7 uses separate, out-of-band signaling channels to carry information needed for advanced features. SS7 has the capability to access databases located in the cellular network for information needed for features such as repeat dialing, call waiting and call return.

PCS—Personal Communications Services

PCS was conceived as a way to provide a low-cost, feature-rich wireless telephone service. Pricing was to be low enough for the service to be affordable to a wide segment of the population. The handsets incorporate two-way paging, short messaging service on the liquid crystal display and voice messaging. Coders and decoders in the handset translate bits sent through the network into characters capable of being displayed on the telephone's display. The following is a quote from the second quarter of the 1996 issue of *Navigator*, a Nortel newsletter. The quote refers to VoiceStream, a Hawaiian PCS cellular network:

> This service, provided over a 100 percent digital network...offers simple, secure, high-quality personal communications in a portable, pocket-sized handset. The VoiceStream network is based on PCS 1900 digital technology, with the multi-functional handset replacing the combination of a mobile phone, pager and answering machine.

PCS offers customers the same features as digital cellular. However, it runs on a different portion of the airwaves, the 1.8 to 1.9 Gigahertz frequencies. In 1993, the FCC announced plans to auction off portions of the 1.8 to 1.9 Gigahertz personal communications services spectrum. Six chunks of spectrum to be auctioned for PCS service were defined as A, B, C, D, E and F blocks. The A and B frequencies were for the 50 MTAs, Major Trading Areas, which are regions that include multiple cities or states. C through F channels were to be in each of the 450 Basic Trading Areas, BTAs. Basic Trading Areas include only one metropolitan area. Bands A, B, and C have 30 MHz of spectrum and bands C, D and E have 10 MHz each. The rules forbid incumbent cellular providers from bidding on frequencies within their own regions. They are not allowed to bid in areas where they have a 20% overlap in the PCS area in their existing cellular coverage.

The idea behind the federal government's promotion of new use of the airways is to encourage competition and raise money for the U.S. Treasury. By dividing up the country into six groups of frequencies, each area could have six personal communications service competitors plus the two existing cellular providers. It is hoped that competition from PCS services will drive prices lower for all cellular service and encourage growth in new wireless services. In fact, in areas where PCS service is available, competition is driving down prices of traditional cellular service. For example, most PCS carriers do not charge a fee for the first minute of incoming calls. In contrast, cellular providers generally charge for the entire length of incoming and outgoing calls. Providing the first minute of incoming calls at no charge is one method of promoting usage. No charge for the first minute of incoming calls encourages PCS users to give out their wireless telephone numbers. Furthermore, because caller ID is included with PCS, callers feel freer about giving out their number because they can simply screen calls based on the identities of incoming callers.

Another creative pricing strategy is the simplification of pricing structures. For example instead of complex pricing, when callers use their cellular phones in other cities, some PCS operators offer roaming at no charge when callers are in cities where the carrier has service. They also offer a flat rate such as 50¢ to 69¢ per call for roaming in other carriers' areas. This is lower and less complicated than traditional cellular roaming

Reducing Fraudulent Use of Wireless Networks

Many cellular telephone users, particularly those in urban areas, have been hit with people making cellular calls on their accounts. The most prevalent technique for stealing airtime is cloning. To clone, thieves use scanners on main thoroughfares to intercept electronic serial numbers and mobile identification numbers transmitted when cellular telephones are turned on. According to Dave Dixon, Vice President of Operations at CTIA, Cellular Telecommunications Industry Association, the cellular fraud reached $650 million annually at its peak. End-users hit by unauthorized use of their service are not liable for charges for stolen airtime fees. Carriers take a loss on these fees.

According to Dave Dixon, cellular fraud has been reduced 40% over the last two years. The reduction was accomplished by use of encryption and authentication. Electronic serial numbers and mobile identification numbers are transmitted by handsets (for billing and routing) whenever they are turned on. These identification numbers are now scrambled in an encrypted format so that cloning is more difficult. In addition, authentication techniques are being implemented in the cellular industry to authenticate that serial numbers are not stolen. Because hackers are so persistent in designing hacking tools to keep up with advances in security techniques, the CTIA is paying labs to try and break industry encryption techniques. The industry hopes to stay ahead of hackers.

fees. Finally, PCS packages such as Sprint's PCS service includes free voice mail, caller ID, call waiting and three-way calling. In some cases, traditional cellular carriers in markets with PCS competition are lowering their prices in response to the presence PCS service. Thus, customers with traditional cellular, as well as PCS services are benefiting from competition.

PCS—Lack of Interoperability and Tower Requirements

Because PCS services operate at higher frequencies than digital and analog cellular, towers need to be spaced more closely together. Higher frequency airwaves are smaller and cannot travel as far as larger, lower frequency airwaves. Therefore, towers need to be spaced more closely together. However, PCS towers are inexpensive to construct. They can be smaller in height and use less power.

In contrast to standardized methods of implementing analog cellular services, PCS services are deployed over incompatible multiplexing schemes. PCS services are developed by multiple manufacturers who use different, incompatible methods of digitizing calls. Wireless handsets using unlike methods of transmitting digital calls cannot be used interchangeably in each carrier's territory. Some carriers purchase equipment that uses one method of digitizing calls and other PCS and cellular providers purchase PCS equipment from manufacturers that use a different, incompatible method of digitizing wireless calls. To complicate matters further, Europe and the rest of the world use still a different, inconsistent method of digitizing and transmitting calls.

Time Division Multiplexing vs. Code Division Multiplexing

All digital wireless services use multiplexing techniques to carry more calls on the same amount of spectrum than analog cellular service. Multiplexers are devices that transmit signals from two or more devices over a single channel. Thus, while all digital services have more capacity than analog cellular, the ways they add capacity are incompatible with each other. Callers cannot necessarily use their cellular telephones when they travel. Personal communications services, PCS, and digital wireless cellular carriers use three ways to multiplex calls. These methods are used on both PCS, 1.8 MHz, and digital cellular, 800 MHz, frequencies.

These different multiplexing schemes are:

- *CDMA*, code division multiplexing used by Sprint PCS, PrimeCo, Bell Atlantic, AirTouch, Ameritech and the country of Korea.

- *TDMA*, implemented by digital cellular providers such as Southwestern Bell Mobility and AT&T Wireless.

- *GSM*, originally Groupe Speciale Mobile, but also known as Global System for Mobile Communications. A standard set by the European Union and used in Europe, the Far East, Israel, much of South America, New Zealand and Australia. The AT&T Wireless TDMA scheme is similar to GSM's TDMA.

Code Division Multiplexing

Code division multiplexing was developed by a U.S. company called Qualcomm. CDMA technology was invented by Qualcomm founder Irwin Jacobs in the late 1980s. The promise of more capacity and a requirement for fewer base stations made by CDMA proponents steered many U.S. digital wireless companies, particularly the Bell telephone companies and Sprint Communications, away from the time division multiple access, TDMA, standard already agreed upon by the TIA, Telecommunications Industry Association, in 1989. The industry has poured billions of dollars into Qualcomm's parent company, Quality Communications. Qualcomm was started in 1985 and introduced code division multiple access in 1989. Service using CDMA was targeted for implementation in 1992. However, technical glitches resulted in actual implementation in 1996.

Code division multiple access is a "spread spectrum" technology. Each conversation transmitted is sent over multiple frequencies. The transmissions are spread among multiple frequencies. This is accomplished through the use of unique 40-bit codes assigned to each telephone or wireless device. Having a unique code assigned to each data or voice transmission allows multiple users to share spectrum or air space. Early proponents of CDMA promised that CDMA would have 20 times the capacity of analog cellular service. Experts differ over the amount of capacity gained with CDMA. Motorola claims eightfold. TDMA gains three to five times more capacity than analog cellular.

In addition to capacity, CDMA handsets use low amounts of power. This can be significant in light of consumer concerns about cellular handsets causing cancer. Lower power emissions translate to less threats of wireless service causing cancer. Finally, calls are transferred, handed-off, from cell to call by a soft handoff method superior to TDMA and analog cellular's handoff. With a soft handoff, a call is rarely dropped during the handoff. For a short period of time during the handoff or transfer, the call is held as it is received and as the cell hands it off. Unfortunately, the decision to use what they perceived as a superior multiplexing method cost many carriers in the U.S. a high price in lost compatibility.

Time Division Multiplexing

The standard that had been agreed upon by the TIA when CDMA was promulgated is time division multiple access, TDMA. In TDMA, the transmission channel is broken into a number of time slots, for example, six. Three of the time slots carry traffic from three devices and three are not used. The three time slots not used ensure that there is no interference between traffic on the time slots carrying traffic. Time division multiplexing has between three and five times the capacity of analog cellular service.

While time division multiplexing is used both in the U.S. and the rest of the world, the methods do not interoperate with each other. The standard in Europe and most of the rest of the world is called GSM, Groupe Special Mobile. The U.S. has settled on a standard called IS-136.

PCS Vendors

The wireless industry is complex. It is made up of organizations formed by mergers, acquisitions and joint ventures. Because massive investment is required to acquire air space for personal communications service, PCS, market the services and build the towers, base stations and switching offices, cable companies, long distance companies and Bell companies have formed joint ventures to offer PCS service. For example, Sprint PCS is a joint venture of three cable companies and Sprint Communications. PrimeCo is a joint venture of Bell telephone companies Bell Atlantic and AirTouch, which bought the Pacific Telesis and U.S. West cellular services. In addition, foreign companies such as Sony Corporation of Japan have investments in personal communications companies such as Next Wave Communications. The joint ventures are a way to share both the expenses and risks associated with the new services. The following three organizations bought the largest amounts of spectrum from the auctions held by the FCC.

PCS PrimeCo, originally a consortium backed by the Baby Bells, is now backed by Bell Atlantic and AirTouch Communications. The consortium was initially supported by RBOCs Bell Atlantic Corporation, NYNEX Corporation, U.S. West and Pacific Telesis. Since the formation of PrimeCo, Pacific Telesis sold its cellular service in April of 1994 to AirTouch Communications. In a further consolidation, U.S. West merged its NewVector domestic wireless business and its interest in PrimeCo into AirTouch Communications. Finally, Bell Atlantic merged with NYNEX. Thus, due to industry consolidations, PCS PrimeCo is now backed by two organizations.

PrimeCo Personal Communications, L.P. operates PCS PrimeCo. It paid $1.1 billion for PCS licenses in 11 Major Trading Areas, MTAs, covering 19 states and 58 million potential customers. It spent another $1 billion by November of 1996 on hardware and software to build its network. It launched service in November of 1996 and as of June 1997 it has PCS offerings in 17 cities. PrimeCo plans to sell telephones and airtime through 34 of its own retail outlets and through other retailers. It sells in outlets such as Radio Shack, Best Buy, and Office Depot.

Sprint PCS is a joint venture of Sprint Communications and cable companies Cox Communications Inc. Comcast Corporation is a cable operator and owner of QVC shopping service and Tele-Communications Inc. Sprint has licenses for PCS spectrum in 35 MTAs covering 65 cities. It plans to purchase licenses in all 50 states. Sprint PCS' strategy is to market to business customers and consumer users. However, its emphasis is more on the business community than PrimeCo's. It reaches residential customers through its distribution by Tandy Corporation's Radio Shack retail chain. Sprint PCS telephones are co-branded with the Sony and Sprint names. Both corporations' names appear on the telephone made for Sprint PCS service.

AT&T Wireless bought McCaw Cellular Communications, Inc. for $12.6 billion in 1993. At the time of the purchase, McCaw Cellular was the largest cellular carrier in the U.S. with service in 105 nationwide markets. According to the Spring 1994 *CTIA Wireless Sourcebook*, the combined population in the areas served by McCaw Cellular was 65,999,231. Interestingly, when McCaw was bought by AT&T, it had never made a profit.

AT&T is unique in that it has a dual strategy of buying an existing cellular network to upgrade and building new PCS services from scratch. Subsequent to the purchase of McCaw, AT&T purchased PCS spectrum from FCC auctions wherever McCaw did not provide service. AT&T has rolled out its own version of digital cellular wireless service in 40 former McCaw service areas. It is based on TDMA. It plans on introducing true PCS in the 1.8 to 1.9 GHz spectrum. It sells dual mode telephones to its customers that will adapt automatically and work in areas with spectrum in traditional cellular and 1.8 to 1.9 GHz PCS frequencies.

Resale of PCS Service—Carriers Selling to Carriers

Not all PCS vendors plan to enter the PCS market by purchasing and building their own networks. MCI for one has plans to resell airtime. It has agreed to purchase ten billion minutes over the next ten years from Next Wave Telecom Inc. MCI wants to offer advanced wireless features to customers without constructing its own network. MCI in a newsletter to telecommunications consultants stated:

> This is not just a resale deal; it allows MCI to interconnect our intelligent long distance and local networks with Next Wave's facilities. This means that MCI can offer advanced network-based features to our wireless customers using Next Wave's licensed spectrum.

Next Wave, which is backed by Sony Corporation of Japan and Pohang Iron & Steel Company of Korea, bought $4.2 billion in spectrum licenses at FCC auctions. Its strategy is to resell airtime to other carriers. It plans to become a carrier's carrier. Unfortunately, as reported in the June 26, 1997 *Wall Street Journal,* Next Wave as well as four other wireless carriers that bid on PCS licenses are asking the FCC to restructure the debt they owe the government for these licenses. The five companies, one of which is bankrupt (Pocket Communications), claim they can't pay what they bid and that the spectrum is worth less than what they bid because there is so much competition for PCS services. An additional 21 companies that won licenses defaulted and the FCC put eighteen of these out for reauction.

Paging Services

Tone only pagers were first introduced in 1956 by Motorola Corporation. Tone only pagers sent a tone to a person's pager. People who were paged called a paging operator or an answering service to find out the telephone number and possibly why a person was trying to reach them. Physicians, plumbers and people who needed to be reached in emergencies were early users of tone only pagers.

Paging sales have boomed ever since they were obtainable in 1956. However, when cellular telephone service was introduced in the late 1980s, many industry experts thought that this would be the death knell for paging because cellular is a two-way service. For example, cellular handset users do not have to run to a telephone every time someone wants to reach them. Industry experts were proved wrong. On the contrary, paging is flourishing. According to Walt Tetschner, president of Tern Systems, a consulting firm in Acton, Massachusetts, "Industry analysts misread the differences between paging and cellular services. I believe there are significantly different needs met in the paging and cellular telephone service market segments. Pagers are always smaller and lower in cost than cellular telephones. The same misreading happened when television was new. Experts predicted that radio was dead. They did not understand the differences between the two media."

How Paging Differs from Wireless Telephone Service

Paging is much cheaper than cellular service; monthly fees range from $8 to $13 and often include usage. Moreover, they are easy to use, convenient and small. A pager fits on a person's belt. Another important advantage of pagers is the length of their battery life. Paging batteries, usually AA, often last for a month or two. PCS batteries last about three days on standby. New digital pagers have a "battery saver" feature where the pager "samples" the air at designated intervals so that the battery is not always in use. This innovation, developed by Motorola Corporation, extends pager battery life.

The question of paging versus mobile telephone service has heated up further with the availability of PCS, personal communications service. PCS is designed to be less costly than traditional cellular service. The handsets used with PCS also incorporate alphanumeric paging and short messaging services. While PCS prices are expected to decline, at present, paging is still lower in cost. The $8 to $13 monthly fee for paging is certainly lower when compared to PCS service with high monthly fees for a set volume of calling or per minute charges for airtime. In Washington, DC, Sprint's most popular package was $40/month in 1996. Although PCS service may take market share from paging services, paging services are adding innovations and growing. Ameritech's 1996 annual report reported growth of paging subscribers by 53% over 1995, as compared to a growth of cellular subscribers of 32.8% from 1995 to 1996.

Advances in Paging Services

Traditional paging services are sustaining growth by marketing, the availability of new spectrum for advanced features and product innovations. In addition to the longer battery life mentioned above, pagers routinely capture the telephone number of the person calling. In contrast to the early tone only pagers, people no longer have to call a third party to find out who paged them. Other innovations include pagers that come in bright colors, pagers that alert people to new messages by vibrating instead of beeping and alphanumeric pagers.

A major advance in paging services is the nationwide reach of some services. A case in point is Mobile Telecommunications Technologies Corporation's SkyTel. SkyTel is an example of a paging service with a nationwide network of towers and satellites that beam signals between states. This is particularly popular with large national companies whose employees travel to remote sites. For example, technical staff such as telecommunications and computer professionals now leave greetings on their voice mail systems telling callers their pager numbers so they can be reached at all times even if they are helping with technical equipment at a remote site. Major customers expect an immediate response to their calls and take it for granted that vendor staff can be reached at all times. This is particularly true in industries such as call centers that have seven-day-a-week, 24-hour-per-day operations.

Staff at all organizational levels have pagers so they can be contacted for repairs, emergencies, important sales opportunities and by family members. Many companies routinely give their salespeople pagers rather than cellular telephones in case a new customer calls in and wants a return call. In this way, people answering the telephone for the sales department are instructed to tell callers that they will get a call back within a specified amount of time.

Nationwide coverage also works well in terms of ease of administration for customers. Customers that manage thousands of pieces of equipment want one point of contact, one bill and large discounts associated with bulk purchases. When a particular vendor provides nationwide coverage, it also has a salesperson, a single customer service number and a single billing contact for major accounts.

When pagers were first sold, they were purchased primarily by business and professional customers. This has changed remarkably. Pagers are now sold in retail outlets. They are status symbols with teenagers and come in bright colors. Motorola estimates that 40% of pagers are used for non-business purposes.

Often, a pager is an adjunct to a cellular telephone. According to a Ram Communications Internet-based white paper posted September 1996, 25% of people with pagers have cellular telephones. A person may be paged via a beeper and return calls from a cellular telephone. In this way, he/she screens calls to avoid paying for incoming nuisance calls. Pagers are more popular in the Far East than in the U.S. In Hong Kong, most people, with the exception of prisoners and very young children, have pagers.

Two-way Paging Using Narrowband PCS

Current two-way pagers allow users to receive pages and short messages and respond to pages with canned messages preprogrammed under buttons of a pager, e.g., "stuck in traffic". Next generation two-way pagers are envisioned incorporating keyboards so that responses to pages can be typed individually. These services could also be used as email vehicles. Two-way paging uses two narrow slices of the PCS spectrum. One channel is used for sending the page and the other for responding to pages. Paging companies are interested in narrowband paging, not only for its two-way capability, but for its ability to provide additional capacity. Non-PCS paging sends out pages using a technique called Simulcast. Simulcast sends each page through every tower in the paging carriers' network. This uses up capacity inefficiently. Narrowband PCS, on the other hand, sends each page only to the same place as the last transmission. This is correct in 90% of the pages. If the pager is not found on the first try, retries are quickly placed through other towers in the carrier's network. In this way, not all of the spectrum is used for each message.

In 1994, the FCC started auctioning off narrowband PCS spectrum. Not all carriers who purchased narrowband spectrum are using it for two-way services. Consider Paging Network Inc. of Dallas, Texas. It is about to offer wireless messaging through its PageNet product name. The wireless messaging service is called VoiceNow. To use VoiceNow, customers will get a pager that is a little bigger than a standard pager. Callers trying to reach a person with a VoiceNow pager will be able to leave a 30-second message. The pager will have the capacity to store about eight 30-second voice messages. It will beep or vibrate when a message is left. The price will be about $12.95 per month.

Wireless for Local Telephone Service

Long distance carriers hoping to provide local calling services are considering wireless options as a way to link homes and businesses to their own central office switches. The biggest hurdle for long distance companies that already have switches in place is the requirement for copper links to individual premises. With wireless service, providers avoid the labor-intensive job of laying cable. Even some Bell telephone companies are considering wireless options as a way to provide local service. Pacific Telesis is looking at wireless as a way to provide service in areas where its fiber or copper facilities are at capacity. It is also looking at wireless as a way to expand into other telephone companies' territories. In particular, Pacific Telesis has entered into agreements to purchase wireless gear for expansion within GTE's territory in California. One option Pacific Telesis is considering is called fixed wireless.

Fixed Wireless

AT&T is testing the use of fixed wireless services to bring local telephone service directly to homes and businesses. Fixed wireless service provides customers with a fixed antenna on their premises. This is an option AT&T is exploring to save money on reaching local customers. AT&T estimates costs of less than $1000 to provide a wireless link. This contrasts with the cost of between $1000 and $2000 to provide a traditional Bell wire-based line. The fixed service AT&T is testing consists of a 13-inch box attached to the outside of a premise which acts as the antenna. Neighborhood antennas beam radio signals to homes. Each neighborhood antenna can serve 2000 homes.

Each 13-inch box provides two telephone lines and a data channel that can be used for Internet connections capable of transmitting at 128 thousand bits per second. This is the same speed of two channels of a BRI ISDN line. Fixed wireless is an example of how wireless service can be deployed for local calling. AT&T's system, however, is not ready for deployment. But, it is one way that AT&T can link callers to its existing switches to provide local and long distance calling.

Debit Cards for Wireless Calling

Debit cards, in conjunction with mobile phones, are being used in parts of the world as a way to provide local telephone service. In a strategy analogous to that used in traditional long distance services, debit cards are being introduced throughout the world for local service. Debit cards are a way to prepay for a certain amount of airtime. Debit cards are particularly popular with people not eligible for credit, such as many in developing countries. In contrast to traditional debit cards where users buy just the debit card, with debit cards made for wireless services, vendors often sell a set amount of usage along with the cellular telephone. For example, in Mexico, Telmex sells cellular debit cards that include a Motorola telephone along with a set number of minutes of long distance.

Specialized Mobile Radio—From Voice to Data

Initially, the specialized mobile radio spectrum, in the 800 to 900 MHz range, was given out for voice applications. Examples of customers who used these services were contractors and concrete companies. They pushed a microphone button to talk. Conversations between the mobile units and from mobile units to a central site were private. However, the specialized networks were not interconnected with anyone else's network. These systems had no interconnection with the public switched telephone network. This spectrum was adapted for new analog data only, in particular, applications by ARDIS and RAM. Eventually, companies such as Nextel and Geotek Communications used the spectrum for digital services to sell voice, data, paging and short messaging services.

Traditional data communications networks transmit data between company sites at fixed locations, a building in New York City, for example, to a building in Chicago. This works well for applications such as payroll, inventory statistics and email, when many employees are based at stationary locations. However, organizations that provide delivery services, field service and installation and transportation services need data communications for mobile employees. Consider companies that send out telephone repair people to fix phone systems. The hours that a technician is on site need to be entered into a computer so that the customer is billed accurately. Other types of field service organizations such as elevator companies, utilities, office equipment suppliers, computer suppliers and transportation services have similar requirements. Giving technicians the capability of entering hours, materials used and causes of repair problems translates to less paperwork for employees and more accurate time and material billing. A technician is less likely to lose track of hours if they are entered at the time of the service call.

Private Networks over Mobile Radio Frequencies

In the late 1970s, companies started using radio frequency networks to transmit data from mobile workers. A prime example of an organization using wireless for private data communications is Federal Express. Federal Express used its private data communications network to give it a major edge over its competitors. The service was used to track the location of packages. Each package had a bar code. When the package was picked up or dropped off for delivery, its bar code was scanned into Federal Express' computer system at the drop-off site or by a driver's hand-held device. The hand-held device transmitted the bar code to Federal Express' computers. The scanning was repeated at each strategic point in the delivery system and transmitted to Federal Express' computers over the radio frequency network. Thus, Federal Express knew where each package was in each step of its journey.

A complementary technology that helped spur the use of radio networks for mobile workers was the availability of portable computers and hand-held devices and scanners for data entry. Improvements in flash memory, developed by Intel, allowed computers to hold information in memory while computers were turned off. However, the information in flash memory could be erased by the user. In addition, lighter batteries and microcircuitry made portable computers and scanners lighter and smaller so that they could be easily carried by technicians and delivery people.

RAM and ARDIS Packetized Radio on Specialized Mobile Radio Spectrum

Rather than develop their own private radio networks, companies often lease specialized mobile radio networks from companies such as RAM Communications and ARDIS. These data-only packet radio systems have been available since 1984. They are used mainly by trucking and dispatch services for communications from mobile workers.

ARDIS is a joint venture of IBM and Motorola Inc. ARDIS' first network facilities were built by Motorola for IBM in the early 1980s for IBM technicians' communications to their offices. RAM Mobile Data USA Partnership is a joint venture between BellSouth Corporation and RAM Broadcasting Corporation of Woodbridge, New Jersey. RAM's first large-scale services were used for electronic mail between people with portable computers.

SMR, specialized mobile radio, uses FCC-allocated spectrum in the 800 to 900 MHz range, close to the spectrum set aside for cellular telephone service. SMR services are two-way. They can be used for sending and receiving data. The speeds on these services are slow. They range from 4.8 Kbps to 19.2 Kbps, which makes them suitable for small bursts of text such as electronic mail, package tracking, license number look-ups by law enforcement agencies and car rental tracking. They were initially analog but carriers are upgrading to digital facilities. An interesting example of a specialized mobile radio carrier upgrading its facilities to digital and expanding its reach of services from data communications over analog facilities to digital voice communications is Nextel Communications, located in Mclean, Virginia.

Specialized Mobile Frequencies for Voice—Nextel

The Nextel wireless telephone offering indicates how new digital wireless telephone services hope to undercut pricing on traditional cellular calling and thus take market share away from these services. Nextel was founded in 1987 and initially offered data communications over analog radio facilities. Nextel's newer wireless telephone service is carried over digital facilities in its 800 to 900 MHz spectrum. The service is used with Motorola telephones and is geared to small- to medium-sized businesses. It does not target residential customers. Motorola Corporation owns 20% of Nextel and the McCaw family, whose own cellular company was purchased by AT&T in 1993, owns 26%.

The Motorola telephones used with Nextel services have a liquid crystal display which can be used for text and numeric paging. Nextel plans to have the telephone service available for 85% of the population by the end of 1998. These areas represent largely metropolitan locations. Vast areas of sparsely populated sections of the country will not be covered. They plan, however, to be accessible from all of the major interstate highways. They won't have towers or service in remote locations with few businesses such as North Dakota and Montana. As of July of 1997, Nextel reached 225 cities, half of its 85% goal. It also has service in nine of the largest international cities such as Shanghai, Vancouver and Mexico City.

One way Nextel hopes to attract customers is by being price-competitive with cellular and PCS carriers. Nextel charges no roaming fees and rounds billing, after the first full minute, to the second rather than to the next full minute. Most other wireless carriers charge in full minute increments which results in higher usage costs. A caller making a 2 1/2 minute call under full minute billing is charged for a full three minutes. In addition, Nextel offers a 30% savings on calls between people on the Nextel service.

This service is not PCS. It is not carried over the 1800 to 1900 frequency range over which true PCS is carried. Unlike PCS carriers, Nextel did not have to participate in costly bidding for new frequencies. It offers wireless telephone service over its pre-existing spectrum. Nextel offers features similar to those of PCS service such as short messaging text paging and voice mail. Because of its proprietary nature, users cannot use their telephones in locations without Nextel service. Nextel claims its service is superior to PCS because it operates in lower frequencies, which because the waves are bigger, they have better penetration into buildings. It also requires fewer towers because it is in a lower frequency where the signal travels further than at higher frequencies.

Another example of a cellular provider which offers voice, data, two-way messaging and paging services over digital specialized mobile radio frequencies is Geotek Communications, Inc. in New Jersey. Geotek was started by an Israeli company as a joint venture with the Armament Development Authority of the State of Israel to commercialize technology developed for military applications. Geotek has voice service in eight cities in the

U.S. It has licenses in the SMR spectrum in 42 Metropolitan Trading Centers. Geotek services are aimed at customers who want to keep track of where their vehicles are. Their system tracks each vehicle through its packet radio data network. This is an interesting example of defense technology transferred to commercial applications.

Building a wireless network takes deep pockets and technical expertise. Nextel is not profitable and Geotek Communications has had technical difficulties. According to its 1997 first quarterly earnings statement, Nextel lost $220 million despite adding 122,600 new mobile units in the quarter. Nextel had bugs in its Motorola equipment when the network was new and Geotek has no interoperability for customers in different cities. A user in one city of Geotek's network cannot use the service when he/she travels to a different city served by Geotek. IBM is helping rescue Geotek by providing Geotek's network with the capability of allowing users in multiple cities to communicate with others over Geotek's network.

Low Earth Orbiting Satellite Networks (LEOS)

Because telephone service is not readily available in much of the developing world, there is an interest in producing a system that can reach all areas of the world without laying miles of cable. In South America, according to Richard A. Keuhn's article in the August 1996 *Business Communications Review*, only 8.5% of the people have telephones. Organizations are raising money and spending large sums on the design of satellite systems that can deliver telephone service, Internet access, facsimile transmission, global paging and data communications in hard-to-reach places with difficult terrain. Few of these systems are in place, but money is being spent on designing the technology, building the systems and obtaining licenses for spectrum.

One technology for delivering these services to all locations worldwide is low earth orbiting satellite networks, LEOs. This development is in contrast to PCS providers whose strategy is to reach areas with high concentrations of population. Traditional satellites, called geosynchronous satellites, orbit at 22,300 miles above the earth's surface. Because they are so high in the sky, the area to which each satellite is capable of beaming signals is large. This is analogous to a flashlight beam. Holding a flashlight higher in the air extends the amount of area the light illuminates. Lowering the flashlight lessens the amount of area the light covers.

However, geosynchronous satellites 22,300 miles above the earth cause a problem with delay on voice and data transmission. The satellites are so high in the sky that there is a delay between when the data is sent and when it is received. Consider voice telephone calls placed in locations where fiber optic cabling is not in place. People who placed calls to international locations in the 1970s and 1980s often experienced clipping and delay. If two people on a telephone call spoke at the same time, parts of words were "clipped" off. The quality of these calls was notably degraded because of delays caused by satellite transmission. The same problem exists in interactive, two-way data communications services.

Low earth orbiting satellites, LEOs, will solve this delay problem because they will be positioned from 435 to 1000 miles from the earth. However, placing satellites lower in orbit lowers the area to which each satellite's transmission reaches. Just as flashlights provide less coverage when they are lower, satellites cover less area when they are not as high in the sky. The requirement for a greater number of satellites increases the cost of LEOs. One service, Teledesic, conceived by Craig McCaw, founder of McCaw Cellular and backed by both McCaw and Microsoft's founder Bill Gates, initially planned to use 840 satellites, but scaled back to 288. Motorola's Iridium project has plans to launch 66 satellites at a cost of $15 million each. This is down from $200 million for each traditional-type, geosynchronous satellite. Motorola Corporation, Teledesic Corporation, Orbcomm Global LP and Odyssey are some of the organizations planning these services.

Vendors of LEOs

Two high-profile backers of LEOs are Motorola Corporation and Craig McCaw, in conjunction with Microsoft Corporation. Other ventures include the Inmarsat Council consortium, the Odyssey cellular network backed by TRW and Teleglobe Canada. A small company with satellite service in place in the U.S. is Orbcomm Global LP, which provides a vehicle tracking service. Fifty percent of Orbcomm is owned by Orbitel Sciences Corporation in Reston, VA. It has other backers in Malaysia and Montreal. Other contenders in this arena are Globalstar, backed by Loral, and ICO Global, backed by Hughes/Comsat.

Motorola Corporation has three LEO projects it is supporting:

Iridium: Iridium is planned to start service in September of 1998 for worldwide wireless telephone with the functionality of PCS, e.g., voice mail, caller ID and call forwarding. Iridium, which was first conceived by Motorola Corporation in 1985, is also planned to be used for paging, data communications and facsimile transmissions. By June of 1997, five of the planned 66 satellites were in place, with seven additional satellites planned within the next month. In May of 1997, Motorola announced an initial public offering of Iridium, although Motorola still owns 25%.

Celestri: Celestri, also backed by Motorola, is positioned as more of a high-speed data and commercial-quality video service than Iridium. In this, it is positioned to compete with Teledesic, which will also offer high-speed data services and teleconferencing.

M-Star: M-Star is also planned for high-speed data applications to be sold to large multi-national corporations.

Teledesic was initially funded by McCaw Cellular. When McCaw Cellular was sold to AT&T Wireless, AT&T Wireless kept a small part of Teledesic. After the sale, Craig McCaw took over the funding of Teledesic and Bill Gates of Microsoft added $10 million. Teledesic's main focus, with its planned 288 LEOs, is to make the Internet accessible from anywhere on earth. The May 27, 1996 *Fortune Magazine* quote from Craig McCaw, the founder of Teledesic, addresses telecommunications needs in the context of the Internet. "The revolution is here. People need high-capacity networks to homes, to remote locations, to farms, to villages that have never had fiber and won't for a long time. It's pretty obvious that there's a need, and that demand will build." Users will only pay for the service for as many minutes as it is used. To connect to Teledesic satellites, users will have to connect an antenna and a signal decoder to their computers or telephones.

Glossary

5ESS   A digital central office manufactured by Lucent Technologies, formerly part of AT&T.

10base-T   IEEE specification for unshielded twisted pair cabling for Ethernet local area networks which transmit at 10 million bits per second. The distance limitation on 10 base-T networks is 100 meters.

100base-T   A proposed standard, compatible with 10base-T for transmitting at 100 megabits over twisted pair cabling on local area networks.

ANI, automatic number identification   The business or residential customer's billing number. Customers such as call centers pay for callers' ANI to be sent to them simultaneously with incoming 800 and 888 calls.

ATM, Asynchronous Transfer Mode   A high-speed switching technique that uses fixed size cells to transmit voice, data and video.

Backbone   A segment of a network used to connect smaller segments of networks together. Backbones carry high concentrations of traffic between smaller segments of networks.

Bandwidth   The measure of the capacity of a communications channel. Analog telephone lines measure capacity in hertz, the difference in the highest and lowest frequency of the channel. Digital channels measure bandwidth in bits per second.

Bit error rate   The percentage of bits received in error in a transmission.

Bps, bits per second   The number of bits sent each second. For example, 1200 bits per second means 1200 bits are sent in one second.

BOC, Bell Operating Company   One of the 22 local Bell telephone companies formerly, before 1984 owned by AT&T. Examples of Bell operating companies are Michigan Bell, Illinois Bell and Pac Bell.

BRI, Basic Rate Interface   The ISDN, integrated services digital network interface made up of two B channels at 64 kilobits each and a signaling channel with a speed of 16 kilobits.

Bridge   A device that connects local or remote networks together. Bridges are used to connect small numbers of networks as bridges do not have routing intelligence. Organizations that wish to connect more than four or five networks use routers.

Broadband   A data transmission scheme where multiple transmissions share a communications path. Cable television uses broadband transmission techniques.

Broadcast   A message from one person or device forwarded to multiple destinations. Voice messaging and email services have broadcast features whereby a user can send the same message to multiple recipients.

CAP, competitive access provider   CAPs originally provided large and medium-sized organizations with connections to long distance providers that bypassed local telephone companies. They now sell local and long distance telephone service.

CCS, common channel signaling   A signaling technique used in public networks. Signals such as those for dial tone and ringing are carried on a separate path from the actual telephone call. CCS allows for telephone company database query used in features such as Caller ID, call forwarding and network-based voice mail. CCS channels are also used for billing and diagnosing public network services.

Centrex   Centrex stands for central exchange. Centrex, like private branch exchanges, routes and switches calls for commercial and non-profit organizations. However, Centrex service is managed by local telephone companies. The computerized Centrex equipment is most often located at a telephone company's central office rather than at a customer premise.

Channels   A path for analog or digital transmission signals. With services such as ISDN, T-1 and T-3, multiple channels share the same one or two pairs of wires.

CIC code, carrier identification code   The four-digit code, (previously three digits,) assigned to each carrier for billing and call routing purposes. AT&T's CIC code is 0288. If someone at a pay telephone dials 1010288 and then the telephone number they are calling, their call is routed over the AT&T network.

CIR, committed information rate   A term used in frame relay networks to indicate the speed of the transmission guaranteed between each customer's site and the frame relay network.

Circuit switching   The establishment, by dialing, of a temporary physical circuit (path), between points. The path, (circuit), is terminated when either end of the connection sends a disconnect signal by hanging up.

CLEC, competitive local exchange carrier   A competitor to local telephone companies that has been granted permission by the state regulatory commission to offer local telephone service. CLECs compete with the Bell or independent telephone company. CLECs are also called ALECs, alternative local exchange carriers or simply local telephone companies.

CO, central office   A local telephone company switch that routes telephone calls. End offices are central offices that connect end-users to the public network.

Compression   Reducing the size of the data, image, voice or video file sent over a telephone line. This lessens the capacity needed to transmit the file.

CPE, customer premise equipment   Equipment such as telephone systems, modems and terminals installed at customer sites.

CSU/DSU, channel service unit/data service unit   A digital interface device that connects customer computers, video equipment and terminals to digital telephone lines.

DCE, data circuit-terminating equipment   A communications device that connects user equipment to telephone lines. Examples are modems for analog lines and CSUs, channel service units for digital lines.

Dedicated line   A telephone line between two or more sites of a private network. Dedicated lines are always available for the exclusive use of the private network at a fixed monthly fee.

DID, direct inward dialing   A feature of local telephone service whereby each person in an organization has his or her own ten-digit telephone number. Calls to DID telephone numbers do not have to be answered by on-site operators. They go directly to the person assigned to the ten-digit DID telephone number.

DMS 100   A digital central office manufactured by Nortel, formerly Northern Telecom.

DS-0, digital signal level 0   The digital signal level 0 is 64 thousand bits per second. It refers to one channel of a T-1, E-1, E-3, T-3, fractional T-1 or fractional T-3 circuit.

DS-1, digital signal level 1   The T-1 transmission rate of 1.54 million bits per second. There are 24 channels associated with DS-1 or T-1.

DS-3, digital signal level 3   Refers to the T-3 transmission rate of 44 million bits per second with 672 channels. (T-3 is equivalent to 28 T-1s.)

DTE, data terminal equipment   Devices that communicate over telephone lines. Examples are multiplexers, PBXs, key systems and personal computers.

E-1   E-1 is the European standard for T-1. E-1 has 2.108 megabits with 30 channels for voice, data or video.

E-3   E-3 is the European standard for T-3. E-3 has a speed of 34.368 megabits, with 480 channels. It is equivalent to 16 E-1 circuits.

Ethernet   A local area network protocol defined by the IEEE. It defines how data is transmitted on and retrieved from local area computer networks.

FDDI, fiber distributed data interface   An ANSI defined protocol whereby computers communicate at 100 million bits per second over fiber-optic cabling. FDDI is used on backbones that connect local area network segments together.

Fiber-optic cable   A type of cable made from glass rather than copper. The key advantage of fiber-optic cabling is that it is non-electric. Thus it is immune from electrical interference and interference from other cables within the same conduit. Fiber-optic cabling can be used for higher-speed transmissions than twisted pair copper cabling.

Firewall   A firewall is software and hardware that prevents unauthorized access to an organization's network files. The intention is to protect files from computer viruses and electronic snooping.

Fractional T-1   Fractional T-1 lines are cheaper and have a fraction of the 24 channel capacity of T-1 lines. The most common capacities are: 2 channels = 128 kilobits; 4 channels = 256 kilobits and 6 channels = 384 kilobits.

Fractional T-3   Fractional T-3 lines have a fraction of the 672 channel capacity of T-3 lines. For example, they might have the capacity of six T-1s or 144 channels. Fractional T-3s are cheaper than a full T-3 line.

Frame relay   Frame relay networks are public data networks commonly used for local area network to local area network communications. Customers connect to frame relay services over telephone lines from each of their locations to the frame relay network. Frame relay services require less maintenance, hardware and upkeep than traditional data communications services for customers with more than about four locations.

Gateway   A gateway device allows equipment with different protocols to communicate with each other. For example, gateways are used when incompatible video systems hold a video conference.

Homepage   A homepage is the default first page of a World Wide Web location that users see when they visit an organization's Web site.

Hub   Each device such as computers and printers on a local area network is wired to the hub, generally located in the wiring closet. Hubs enable local area networks to use twisted pair cabling rather than more expensive, harder to install and move coaxial cabling. Hubs are sometimes referred to as concentrators.

IC, interexchange carrier   Interexchange carriers are the long distance companies that sell toll free 800, international and outgoing telephone service on an interstate basis. With the passage of the Telecommunications Act of 1986, local Bell telephone companies will start to sell interstate telephone service when they meet FCC mandated guidelines for connecting competitors to their networks.

ILEC, incumbent local exchange carrier ILECs refer to the Bell and independent telephone companies that sell local telephone service. This term differentiates telephone companies that were the providers of telephone service prior to the Telecommunications Act of 1996 and new competitors such as MSF, Teleport, MCI and AT&T.

Internet The Internet, with a capital I, is composed of multiple networks tied together by a common protocol, TCP/IP.

Intranet An intranet is the use of World Wide Web technologies for internal operations. Intranets are used by organizations as a way to make corporate information readily accessible by employees. An example is a corporate telephone directory accessed by a browser.

IP, Internet protocol The part of TCP/IP, the Internet protocol that performs the addressing function for networks. Each device on an Internet network is assigned a 32 bit address. The Internet is running out of addresses and standards bodies are reviewing ways to upgrade the address schemes so that more addresses will be available.

ISDN, integrated services digital network ISDN is a standard digital network that lets users send voice, data and video over one telephone line from a common network interface.

ISP, Internet service provider An Internet service provider connects end-users to the Internet via telephone lines. The ISP has banks of modems and devices such as ISDN interfaces for its own customers to dial into. The ISP then rents telephone lines to the Internet from its own location. Some Internet service providers such as UUNet also own Internet backbone networks.

Key system Key systems are on-site telephone systems geared to the under 100 telephone organizations. Like PBXs, they switch calls to and from the public network and within user's premises.

LAN, local area network A local area network is located on an individual organization's premise. It enables computer devices such as personal computers, printers, alarm systems and scanners to communicate with each other. Moreover, LANs allow multiple devices to share and have access to expensive peripherals such as printers, fax servers, modem servers and centralized databases.

LATA, local access transport area At divestiture in 1984, LATAs were set up as the areas in which Bell telephone companies were allowed to sell local telephone services. LATAs cover metropolitan statistical areas based on population sizes. For example, Massachusetts has two LATAs and Wisconsin has four LATAs but Wyoming, which has a small population, has one LATA. The rules of divestiture decreed that long distance telephone companies such as AT&T, Sprint and MCI were allowed to carry calls between LATAs but that Bell telephone companies such as Illinois Bell could carry calls only within a LATA.

Leased line   A leased line is analogous to two tin cans and a string between two or more sites. Organizations that rent leased lines pay a fixed monthly fee for the leased lines that are available exclusively to the organization that leases them. Leased lines can be used to transmit voice, data or video. They are also called private and dedicated lines.

LEC, local exchange carrier   Any company authorized by the state public utility commission to sell local telephone service.

Local loop   The local loop is the telephone line that runs from the local telephone company to the end user's premise. The local loop can be made up of fiber, copper or wireless media.

MAN, metropolitan area network   A metropolitan area network is a network that covers a metropolitan area such as a portion of a city. Hospitals, universities, municipalities and large corporations often have telephone lines running between sites within a city or suburban area.

Mbps, million bits per second   A transmission speed at the rate of millions of bits in one second. Digital telephone lines measure their capacity or bandwidth in bits per second.

Multiplexing, or mux   Multiplexing is a technique whereby multiple devices can share a telephone line. With multiplexing, users do not have to lease individual telephones for each computer that wishes to communicate. T-1 multiplexers enable 24 devices to share one telephone line.

Network   A network is an arrangement of devices that can communicate with each other. An example of a network is the public switched telephone network over which residential and commercial telephones and modems communicate with each other.

NT1, network termination type 1   The device between an ISDN line and an ISDN terminal adapter. The NT1 plugs into the ISDN jack. It provides a point where the network provider can test the ISDN line. The NT1 also converts the ISDN line from the telephone company's two-wire to four-wire cabling. The four wires are the "inside" portion of the cabling.

Packet switching   A network that routes data in units called packets. Each packet contains addressing and error-checking bits as well as transmitted, user data. Packets can be routed individually through a network such as an X.25 network and be assembled at the end destination.

PBX, private branch exchange   PBXs are computerized on site telephone systems located at commercial and non-profit organizations' premises. They route calls both within an organization, and from the outside world to people within the organization.

PCMCIA, portable computer memory card industry association   An industry group that has developed a standard for peripheral cards for portable computers. PCMCIA cards are used for functions such as modems and for additional memory.

POP, point of presence   A POP refers to a long distance company's switch that is connected to the local telephone company's central office. The POP is the point at which telephone and data calls are handed off between local telephone companies and long distance telephone companies.

POTs, plain old telephone lines   Telephone lines connected to most residential and small business users. POTs lines are analog from the end user to the nearest local telephone company equipment. People using POTs service for data communications with modems are limited in the speed at which they can transmit data.

PRI, primary rate interface   PRI is a form of ISDN, integrated services digital network, with 23 paths for voice, video and data and one channel for signals. Each channel has the capacity of 64 kilobits per second.

RBOC, Regional Bell Operating Company   At divestiture, in 1984, the Justice Department organized the previous 22 bell telephone companies into seven Regional Bell Operating Companies. Examples of RBOCs are Ameritech and BellSouth. Four of the RBOCs have merged. Pacific Telesis merged with SBC, and NYNEX merged with Bell Atlantic. There are now five RBOCs. Before divestiture, all of the Bell telephone companies were owned by AT&T.

Router   A device, with routing intelligence, that connects parts of local and remote networks together. Because they use routing tables to look up addresses for each message, routers introduce delays into networks.

Server   A server is a specialized shared computer on the local area network with corporate files such as electronic mail. It can also be used to handle sharing of printers, fax machines and groups of modems.

SMTP, simple mail transfer protocol   The electronic mail protocol portion of the TCP/IP protocol used on the Internet. Having an electronic mail standard that users adhere to enables people on diverse local area networks to send each other email.

SONET, synchronous optical network   A standard for transmitting high-speed digital bits over fiber optic cabling. Telephone companies use SONET to transmit data from multiple customers.

Switched 56   A digital "dial-up" data communications service. It is used for data communications devices that communicate less than two to three hours a day. If ISDN is not available, switched 56 services can be used for video conferencing.

T-1   A North American and Japanese standard for communicating at 1.54 million bits per second. A T-1 line has the capacity for 24 voice or data channels.

T-3   A North American standard for communicating at speeds of 44 million bits per second. T-3 lines have 672 channels for voice and/or data. Fiber optic cabling or digital microwave is required for T-3 transmissions.

TA, terminal adapter A terminal adapter is used with ISDN service. The terminal adapter allows multiple voice and/or data devices to share a digital ISDN line. The terminal adapter sits between the data communicating device or the telephone and the ISDN line.

TCP/IP, transmission control protocol/Internet protocol The set of protocols used in the Internet and also by organizations for communications between multiple networks.

UTP, unshielded twisted pair Most inside telephones and computers are connected together via unshielded twisted pair copper cabling. The twists in the copper cables cut down on the electrical interference of signals carried on pairs of wire near each other and near electrical equipment.

WAN, wide area network Wide area networks connect computers that are located in different cities, states and countries.

WWW, World Wide Web The World Wide Web has both multimedia capabilities. It links users from one network to another when they "click" on highlighted text. It was developed in 1989 to make information on the Internet more accessible.

X.25 An ITU defined packet switching protocol for communications between end-users and public data networks. X.25 is slower and older than frame relay service.

Bibliography

AUERBACH, JON, "Sorry Wrong Phone," *The Boston Globe*, July 12, 1997, pp. 1 and 10.

BRAY, HIAWATHA, "From Sound of Things, BBN Helped Pioneer 'Net,' *The Boston Globe*, May 7, 1997, sec. D, p.2.

CANLEY, LESLIE, "Why Phone Rivals Can't Get into Some Towns," *TheWall Street Journal*, August 19, 1996, pp. B1 and B3.

COLE, CAROLINE LOUISE, "Cable PCs: Speed at Cost of Privacy," *The Boston Globe*, June 8, 1997, pp. 1W and 11W.

CORTESE, AMY, *Business Week*/Harris Poll, "A Census in Cyberspace," *Business Week*, p. 84, May 5, 1997.

ENG, PAUL M., "Cybergiants See the Future—and It's Jack and Jill," *Business Week*, April 14, 1997, p. 44.

FLINT, ANTHONY, "Skin Trade Spreading Across U.S." *The Boston Sunday Globe*, December 1, 1996, pp. A1 and A36.

GAFFIN, ADAM, "Net Pioneers See No End to Their Grand Experiment," *Network World*, August 22, 1997, pp. 1 and 61.

GRULEY, BRYAN AND QUENTIN HARDY, "Wireless Bidders Ask to Restructure Debt," *The Wall Street Journal*, June 26, 1997, pp. A3 and A10.

HIGGINS, RON, "Internet May Suffer as U.S. Telcos Eye Global Markets," *Network World*, March 24, 1997, p.48.

HOF, ROBERT D. AND ELSTROM BROWDER, "Special Report: Internet Communities," *Business Week*, May 5, 1997, pp. 64–80.

_____, "Netspeed at Netscape," *Business Week*, February 10, 1997, pp.79–86.

Just the Facts, p.2, Corning Incorporated, July, 1995.

KUEHN, RICHARD A., "Ruminations from Rio," *Business Communications Review*, August 1996, pp. 76–74.

KUPFER, ANDREW, "Craig McCaw Sees an Internet in the Making," p. 67, *Fortune*, May 27, 1996.

LITTWIN, ANGELA, "ADSL: Ready for Prime Time?" *Telecommunications*, December 1996, pp. 35–44.

MACLURE, MITTLEMAN, "Cautions about Car Telephones and Collisions, *The New England Journal of Medicine*, vol. 336, February 13, 1997, pp. 501–502.

MADSEN, HUNTER, "Reclaim the Deadzone," *Wired Magazine*, p. 206, December 1996.

MCSHANE, TIM, "56K bit/sec Modems Offer Promise, but Pitfalls, Too," *Network World*, April 14, 1997, p. 45.

NAIK, GAUTAM, "Baby Bells Profit by Tapping Phone Paranoia," p. B1, *The Wall Street Journal*, September 3, 1996.

NEWMAN, NATHAN, "How Netscape Stole the Web, or Destroying the Village in Order to Save It," *Enode,* vol. 2, no. 2, February 1997.

"On-Line Advertising Revenues Rise," Technology Briefs, *The Wall Street Journal*, Interactive Edition, June 16, 1997.

QUICK, REBECCA, "Internet Domain: Cyberspace Expands with New Addresses," *The Wall Street Journal*, February 6, 1997, p. B6.

REDELMEIER, TIBSHIRANI, "Association between Cellular-Telephone Calls and Motor Vehicle Collisions," *The New England Journal of Medicine*, vol. 336, February 13, 1997, pp. 453–458.

TROWT-BAYARD, TOBY, *Videoconferencing, the Whole Picture*, p. 47, New York, NY: Flatiron Publishing, Inc., 1994.

WEBER, THOMAS E., "For Those Who Scoff at Internet Commerce, Here's a Hot Market," *The Wall Street Journal*, May 20, 1997, pp. A1 and A8.

ZIEGLER, BART, "In the Net," *The Wall Street Journal*, November 18, 1996, p. R21.

Index